**INTRODUCTION
TO RELIABILITY
IN DESIGN**

McGRAW-HILL BOOK COMPANY

New York
St. Louis
San Francisco
Auckland
Düsseldorf
Johannesburg
Kuala Lumpur
London
Mexico
Montreal
New Delhi
Panama
Paris
São Paulo
Singapore
Sydney
Tokyo
Toronto

CHARLES O. SMITH
Professor of Engineering
University of Detroit

Introduction to Reliability in Design

This book was set in Times Roman by Bi-Comp, Incorporated.
The editors were B. J. Clark and Madelaine Eichberg;
the cover was designed by Chris Kristiansen;
the production supervisor was Thomas J. LoPinto.
The drawings were done by ECL Art Associates, Inc.
Kingsport Press, Inc., was printer and binder.

Library of Congress Cataloging in Publication Data

Smith, Charles O.   date
    Introduction to reliability in design.

    1. Reliability (Engineering)   2. Engineering
design.   I. Title.
TS173.S6   620'.004'5   75-44421
ISBN 0-07-059083-4

**INTRODUCTION
TO RELIABILITY
IN DESIGN**

Copyright © 1976 by McGraw-Hill, Inc.   All rights reserved.
Printed in the United States of America.   No part of this publication
may be reproduced, stored in a retrieval system, or transmitted, in any
form or by any means, electronic, mechanical, photocopying, recording, or
otherwise, without the prior written permission of the publisher.

1 2 3 4 5 6 7 8 9 0 KPKP 7 8 3 2 1 0 9 8 7 6

See Acknowledgments on pages xiii–xiv.   Copyrights included on this page by reference.

CONTENTS

	Preface	xi
	Acknowledgments	xiii
Part	**One**	
Chapter 1	**Introduction**	**3**
1-1	What is reliability?	3
1-2	Probability in engineering	5
1-3	Adequate performance	5
1-4	The tasks of reliability	6
Chapter 2	**Reliability Function**	**8**
2-1	Probability and reliability	8
2-2	Instantaneous failure rate	9
2-3	General reliability function	10
2-4	Graphical representation of reliability	11
2-5	Mortality curve	12
2-6	Expanded reliability function	14

Chapter 3 Component Life — 15

- 3-1 Component mortality — 15
- 3-2 Mean time to failure (MTTF) — 16
- 3-3 Mean time between failures (MTBF) — 16
- 3-4 A priori–a posteriori — 17
- 3-5 Useful life — 18
- 3-6 Wearout — 20
- 3-7 Early life — 23
- 3-8 Expanded reliability function — 25

Chapter 4 Series and Parallel Systems — 29

- 4-1 Introduction — 29
- 4-2 Series systems — 31
- 4-3 Energize–de-energize — 32
- 4-4 Parallel systems — 34
- 4-5 Combined series-parallel systems — 36
- 4-6 Block diagrams (or logic diagrams) — 38
- 4-7 Binomial distribution — 38
- 4-8 Poisson distribution — 41
- 4-9 Multinomial distribution — 42
- 4-10 Solution to 4-6 — 44

Chapter 5 Standby Systems — 50

- 5-1 Introduction — 50
- 5-2 Standby model — 51
- 5-3 Two-unit standby system — 52
- 5-4 Multiunit standby system — 55
- 5-5 Some summary comments — 57

Chapter 6 Conditional Probability — 63

- 6-1 Introduction — 63
- 6-2 Dual function — 63
- 6-3 Bayes' theorem — 66

Chapter 7 Multimode Function — 74

- 7-1 Introduction — 74
- 7-2 One possible procedure — 75

Chapter 8 Derating and Maintenance — 79

- 8-1 Derating — 79
- 8-2 Maintenance — 84

Chapter 9 Reliability Testing — 92

- 9-1 Introduction — 92
- 9-2 Testing program — 94

9-3	Parameter estimation	102
9-4	Accelerated testing	105
9-5	Sequential testing	111

Chapter 10 Broad Guidelines for Design 117

10-1	Introduction	117
10-2	Approaches to the problem	117
10-3	Loading	118
10-4	Environment	118
10-5	Material behavior	119
10-6	Three broad steps	119
10-7	Insuring reliability	119
10-8	Design checklist	121
10-9	Specifications and tolerances	122
10-10	Human factors	122
10-11	Design review	123
10-12	How much reliability?	124
10-13	Summary comment	127

Part Two

Chapter 11 A Possible Timesaver 131

| 11-1 | The problem | 131 |
| 11-2 | One solution | 132 |

Chapter 12 Failure Rates and Design 139

12-1	The problem	139
12-2	Failure rate estimation	139
12-3	Calculated versus actual failure rates	144
12-4	Prior experience	145

Chapter 13 Project Mercury 146

13-1	The situation	146
13-2	The Project Mercury approach	147
13-3	Comments	150

Chapter 14 Design of a DC-DC Converter 152

| 14-1 | The problem | 152 |
| 14-2 | One working design concept | 153 |

Chapter 15 Availability Concept 156

| 15-1 | The problem | 156 |
| 15-2 | One solution | 158 |

Chapter 16 Human Reliability — 162
- 16-1 The problem — 162
- 16-2 One solution — 163

Chapter 17 Super-Reliability — 167
- 17-1 The problem — 167
- 17-2 Possible solutions — 168

Chapter 18 Safety Factor and Reliability — 172
- 18-1 The problem — 172
- 18-2 A solution — 174

Chapter 19 Reliability Allocation — 181
- 19-1 The problem — 181
- 19-2 One solution — 186
- 19-3 Redundancy allocation — 188

Chapter 20 Some Problems to Wrestle with — 196
- 20-1 Gas Turbine engine reliability — 196
- 20-2 Design adequacy assessment — 197
- 20-3 Accelerated fatigue test — 197
- 20-4 Snap-fit assemblies in polymers — 197
- 20-5 Steering knuckle redesign — 198
- 20-6 Structural criteria for composite materials — 200
- 20-7 Reliability in software systems — 201
- 20-8 Service life evaluation — 201
- 20-9 Deployment of a space antenna — 201
- 20-10 Reliability and ESP — 202
- 20-11 Plastic integrated circuits — 202

Appendix A Notes on Probability — 208
- A-1 Introduction — 208
- A-2 Definitions — 209
- A-3 Classical definition of probability — 209
- A-4 Relative frequency definition of probability — 209
- A-5 Fundamental laws of probability — 211
- A-6 Combinations and permutations — 215
- A-7 Random variables — 217
- A-8 Distribution functions — 218
- A-9 Parameters of a distribution — 219
- A-10 Probability distributions — 221

Appendix B Equipment and Factors in Reliability Testing — 227
- B-1 Generalities — 227
- B-2 Why simulate reality? — 228

B-3	Temperature environments	228
B-4	Pressure environments	229
B-5	Corrosive/erosive environments	230
B-6	Mechanical environments	232
B-7	Combined environments	234
B-8	Test sequence	235
Appendix C	**Concept of Ranking**	**236**
Appendix D	**Algebra of Normal Distribution Functions**	**243**
Appendix E	**Use of the Weibull Distribution**	**245**
E-1	Data Evaluation	245
E-2	**Comparison of Two Sets of Test Statistics**	**250**
Appendix F	**The Cumulative Standard Normal Distribution Function**	**257**
Appendix G	**Values of the Negative Exponential**	**258**
	Index	**265**

PREFACE

This book is primarily intended for use in a first course in Reliability at the Senior-Graduate level, but I believe it can also be of substantial value and use to any practicing engineer who desires to gain a basic understanding of reliability, especially with relation to design. The book has two parts, differing substantially in approach and format.

    Part I consists of ten chapters covering definition of reliability, the general reliability function, and basic concepts. The fundamental aspects of series, parallel, standby, and mixed systems are treated. Conditional probability and multimode function are used as additional methods of approach to problem solution. Consideration is given to derating and reliability testing. Part I concludes with a set of broad guidelines for designing reliability into a given situation and a discussion of factors involved in deciding how much reliability is appropriate. A number of relatively simple problems are included, both as part of the text material and at the end of each chapter to permit development of some rudimentary skill for dealing with reliability problems.

    Part II consists of a series of specific examples taken from published literature. It must be recognized that inclusion of these examples does not imply the solutions are the best, or only, ones. Neither does inclusion imply use of the latest techniques. The examples only show how real people dealt with real problems. Obviously, continued improvement is a common and appropriate goal. The concepts and approaches illustrate things that can be done and, it is hoped, will stimulate ideas on ways to approach problems.

Appreciation is expressed to the students at the University of Detroit who have patiently (and maybe gleefully?) pointed out places of possible confusion in the notebook predecessor of this text. The students have also brought to my attention a number of specific examples of reliability application for our mutual benefit. I take responsibility for any errors still remaining. I further appreciate the many hours of expert typing by Mrs. Imogene Marshall of the University of Detroit.

CHARLES O. SMITH

ACKNOWLEDGMENTS

AMERICAN INSTITUTE OF AERONAUTICS AND ASTRONAUTICS for permission to use part of Chap. 15 from Carpenter, Safety Assurance for Extended Manned Missions, *Journal of Spacecraft and Rockets*, vol. 4, no. 4 (April 1967), pp. 448–451.

AMERICAN SOCIETY OF MECHANICAL ENGINEERS for permission to use part of Chap. 15 from Carpenter, Space-Manned Interplanetary Travel, *Mechanical Engineering* (June 1966), pp. 44–48.

AMERICAN STATISTICAL ASSOCIATION for permission to use Table 9-1 from Epstein and Sobel, Life Testing, *Journal of the American Statistical Association*, vol. 48, no. 263 (Sept. 1953).

COMPUTER DESIGN for permission to use material in Chap. 17 from Roberts, Increasing Reliability of Digital Computers, *Computer Design*, vol. 8, no. 1 (Jan. 1969), pp. 44–48.

ELSEVIER SCIENTIFIC PUBLISHING COMPANY, AMSTERDAM, for permission to use Figs. E-6, E-7, E-8, E-9, E-10, and E-11 from Johnson, "Statistical Treatment of Fatigue Experiments," 1964.

HARCOURT BRACE JOVANOVICH, INC. for permission to use Probs. 3-12 and 5-7 from Beckmann, "Probability in Communication Engineering," 1967.

INSTITUTE OF ELECTRICAL AND ELECTRONIC ENGINEERS for permission to use material in Chap. 11 from Jensen and Bellmore, An Algorithm to Determine the Reliability of a Complex System, *IEEE Transactions on Reliability*, vol. R-18, no. 4 (Nov. 1969), pp.

169–174, material in Chap. 12 from Tanner, Operational Reliability of Components in Selected Systems, *1966 Symposium on Reliability,* and material for the basis of Chap. 20 (except Sec. 20-5) from *Proceedings of the Annual Reliability and Maintainability Symposium, 1972–1975.*

MCGRAW-HILL BOOK COMPANY for permission to use Example 4-9, Probs. 4-17 and 4-30, and Example 6-3 from Roberts, "Mathematical Methods in Reliability Engineering," 1964, Tables C-1 and C-2, Figs. E-1, E-2, and E-3 from Lipson and Sheth, "Engineering Experiments," 1973, and Appendix G from Amstadter, "Reliability Mathematics," 1971.

OPERATIONS RESEARCH SOCIETY OF AMERICA for permission to use part of Chap. 19 from Kettele, Least-Cost Allocations of Reliability Investment, *Operations Research,* vol. 10 (March-April 1962), pp. 249–265.

PERGAMON PRESS LTD. for permission to use Examples 4-2 and 4-3, and Probs. 4-7 and 4-19 from Dummer and Griffin, "Electronics Reliability—Calculation & Design," 1966.

PRENTICE-HALL, INC. for permission to use Example 8-1, and Probs. 5-8, 6-11, and 8-3 from Bazovsky, "Reliability Theory and Practice," 1961, Probs. 3-17 and 4-2 from Miller and Freund, "Probability & Statistics for Engineers," 1965, and Example 4-6 and Prob. 4-16 from Arinc, "Reliability Engineering," 1964.

SOCIETY OF AUTOMOTIVE ENGINEERS for permission to use material for the basis of Sec. 20-5 from SAE Papers 690500 and 690501.

SPARTAN BOOKS for permission to use material in Chap. 16 from Pontecorvo, Method of Predicting Human Reliability, "Annals of Reliability & Maintainability," vol. 4, 1965, pp. 337–342.

JOHN WILEY & SONS, INC. for permission to use Examples 6-1 and 6-2 and Probs. 6-1, 6-2, and 6-3 from Breipohl, "Probabilistic System Analysis," 1970 and Appendix F from Hald, "Statistical Tables and Formulas," 1952.

The author also extends his appreciation to Profs. E. B. Haugen and D. B. Kececioglu for originally arousing his interest in reliability at a National Science Foundation Institute at the University of Arizona in 1966. Further acknowledgment is due Dr. Kececioglu for ideas which form bases for a number of the problems in the early chapters of this book.

PART ONE

1
INTRODUCTION

1-1 WHAT IS RELIABILITY?

Reliability is a measure of the capacity of a piece of equipment to operate without failure when put into service. Reliability has been defined in various ways. One of the best is that of the National Aeronautics and Space Administration (NASA) which defines reliability as *the probability of a device performing adequately for the period of time intended under the operating conditions encountered.* Reliability is always a probability associated with a no-failure performance of a device (up to and including a large system) after an accumulated time (can be a very short or a very long period) in a specific environment over a given period of operation with some desired level of confidence.

Reliability is, therefore, the probability that a given system will perform as anticipated. In the case of a rocket with its complex system of thousands of parts during countdown on the launching pad, reliability is the probability that the mission will not abort for any reason, including such things as a tiny, loose bead of weldment ready to be vibrated into a microcircuit. In a rather real sense, anything which is reliable is, by definition, well made.

Since reliability is defined as a probability, its calculation is one form of applied mathematics. Probabilistic and statistical methods, however, just like any other mathematics, are aids to, not substitutes for, logical reasoning. A blind statistical approach, without considering the dictates of common sense, will produce faulty or inaccurate conclusions on the bases of apparently valid methods of dealing with seemingly legitimate data. The validity of a statistical result is no better than the input data and the manner in which they are used.

Reliability is identified with a state of knowledge, not a state of things. To demonstrate, consider a fictitious example. Imagine an astronaut perched in his capsule in the third stage of a rocket. Just as the engineer is about to press the button which fires the first stage, the astronaut asks, "Is this system reliable?" The countdown stops while the engineer answers, "Yes, it is." The astronaut asks, "How do you know?" "Because we fired sixteen of them and the last fourteen worked." The astronaut then asks, "Have you fired *this one?*" "No. If we had, you wouldn't be in it." "Then how do I know this one is reliable?" The engineer replies, "Swing your periscope around and look over the firing range. We have ten more rockets just like yours over there. We'll fire them for you." While the astronaut watches, the first rocket is fired and blows up on the pad. Five more are fired and fly properly. Obviously, after the first failure, both the engineer and the astronaut would have regarded the rocket with the astronaut as less reliable. If the other nine were fired without mishap, both men would have been justified in considering that the reliability had increased. *What has changed is knowledge of the system. The original rocket with the astronaut has not changed for it is still the same system it was when countdown started.* The reliability of the rocket depends on the results of the test firings because we believe that information obtained from the tests is relevant to inferences drawn about the astronaut's rocket. Reliability is a subjective concept, not an objective one. In reliability, we are concerned with a minimally prejudiced description of what is known, maintaining certainty only about what is known, and being maximally vague about what is unknown.

Reliability cannot be used to predict discrete events; only probabilities or averages are predicted. Reliability will not predict that a device will operate for a given number of hours before failing, e.g., a mechanical clock running down. We can predict the probability that a device will operate for a specified number of hours, *or* that, on an average, some number of failures will occur in a given period of time, *or* that there will be a certain average time between failures. A given device might fail immediately after being put into operation or it might function properly for an abnormally long period of time. Reliability will not predict which. Reliability is well described as a general statement of what is expected to happen.

The terms (*predicted, inherent,* and *demonstrated*) for classifying reliability are almost self-explanatory. Predicted reliability represents performance expected of functioning equipment after the design has undergone all corrective actions and has been established as appropriate for use. Predicted reliability can be established by design calculations before assembly or test of the final system. Inherent reliability results from intrinsic values of function and design and can change only with changes in design or degradation during operation.

Demonstrated reliability (determined from test and/or field performance) is subject to both time and economic constraints. Demonstrated reliability is seldom firmly established with a new system but is expected to improve with time and operation and ultimately approach inherent reliability.

1-2 PROBABILITY IN ENGINEERING

The word *probability* is often accepted rather skeptically in engineering practice, which holds itself to be an "exact science" despite much use of empiricism. Engineering, in one sense, is a form of applied physics. Many physical phenomena can be described only by using the theory of probability. Examples are the laws of decay of radioactive substances and statistical mechanics which describe changes of state of atoms and basic particles. Probabilistic calculations thus have real meaning for engineers. We should not be surprised to discover that equipment failures which result from interactions of heat, electric and magnetic fields, "static" loadings, and vibrations can best be described in probabilistic terms.

Directly, or indirectly, probability plays an important part in all problems of engineering, science, business, and individual daily lives which involve an element of uncertainty. Probability serves as a substitute for certainty. There are very few situations in which we have complete information, i.e., all relevant facts. Therefore, we have no choice but to generalize from samples and we cannot be absolutely certain of the conclusions we may reach. Probability for the occurrence of an event refers to what will happen in the long run in a large number of similar events. The fact that probability statements cannot be proved either right or wrong on the basis of occurrence (or nonoccurrence) of single events implies neither that the statements were in error nor that we should proceed to make wild predictions in terms of probabilities. Probability is a measure of what is expected to happen on the average if the given event is repeated a large number of times under identical conditions. A discussion of probability concepts and basic mathematical operations is given in Appendix A.

1-3 ADEQUATE PERFORMANCE

The definition of reliability includes adequate performance of the given device. There is no general definition of adequate performance. Criteria for "adequate performance" must be carefully and exactly detailed, in advance, for each system considered. For example, performance which is adequate for a tire on a private automobile would be woefully inadequate for a tire on a car running in the Indianapolis "500." Careful detailing of the function (excerpted exactly from performance specifications or in terms of required performance) which the device must perform in operation is absolutely imperative. Only under these circumstances will all cases of failure be caught. Only in such cases can criteria of performance be specified. Measurement of all important performance parameters is mandatory. As long as the parameters remain within the specified limits, the device is judged to be performing adequately. When one (or more) performance parameter drifts out of the specified limits of tolerance, the device is judged to have failed or malfunctioned. *The precise definition of adequate performance is completely dependent on the case under consideration.*

1-4 THE TASKS OF RELIABILITY

In Sec. 1-1, there was a qualitative definition of reliability. We can quantitatively define reliability as the probability that the device will meet the qualitative definition: If T is the time to failure of the device (T itself is a variable), then the probability that it will not fail in the given environment before time t, i.e., its reliability, is

$$R(t) = P(T > t) \qquad (1\text{-}1)$$

The device is assumed to be working properly at time $t = 0$. No device can work forever without failure. These statements are expressed as

$$R(0) = 1 \qquad R(\infty) = 0 \qquad (1\text{-}2)$$

and $R(t)$ is a nonincreasing (at best, constant; generally decreasing) function between these two extremes. For t less than zero, reliability has no meaning. We define it as unity, however, since the probability of failure for $t < 0$ is zero.

The first task of reliability engineering is to derive and investigate Eq. (1-1). This is done first for a single component. The reliability of an entire system is then determined in keeping with the configuration and functions of the components composing the system. This system, in turn, may be a subsystem of a more complex system. This building process continues until the complete system under consideration has been treated. The complete system may be so complex as to include entire organizations, such as maintenance and repair groups with their personnel.

When the properties and interdependencies of the system reliability [Eq. (1-1)] are understood, the second task of reliability engineering is to find the best way of increasing the reliability. The most important methods for doing this are:

1 Reduce the complexity of the system to the minimum essential for the required operation. Nonessential components and unnecessarily complex configurations only increase the probability of system failure.
2 Increase the reliability of the components in the system.
3 Parallel redundancy: One or more ("hot") spares operate in parallel. If one fails, others still function.

4 Standby redundancy: A ("cold") spare is switched in to take over the function of a component or subsystem that has failed.

5 Repair maintenance: Failed components are replaced by a technician rather than switched in as in (4). Replacement is neither automatic nor necessarily immediate.

6 Preventive maintenance: Components are replaced by new ones periodically even though they may not have failed by the time of replacement.

Combinations of these methods are possible. For example, a high degree of reliability might be achieved by repairing failed standby units while the system is operating on others, i.e., a combination of methods 4 and 5.

Methods 1 and 2 are obviously limited by the current state of technology. (In the entire discussion which follows, we shall assume these limits are achieved.) The other four methods enable us, in principle, to develop systems having reliabilities which approach 100 percent for all arbitrarily chosen mission times t. It is not always feasible to do this in view of constraints such as limits on weight, space, cost, or maintainability and availability. (Maintainability of a device is defined as the probability of repairing the device in a given system in a given time. Availability of a device is defined as the probability that the device is operating satisfactorily at any point in time where the total time includes operating time, active repair time, logistic time, and administrative time.)

The third task of reliability engineering is to maximize system reliability for a given weight, size, or cost. Conversely, the task may be to minimize weight, size, cost or other constraints for a given reliability.

PROBLEMS

1-1 Read "The Deacon's Masterpiece" by O. W. Holmes. Prepare a commentary on this as an example of reliability in design.

1-2 What is the distinction between reliability and quality assurance?

2

RELIABILITY FUNCTION

2-1 PROBABILITY AND RELIABILITY

Consider a test in which a large number of trials are made. Each trial has an equal chance of resulting in event A. If A does not result, however, then event B does. The probability of occurrence of event A is defined as the ratio of the number of trials from which A results to the total number of trials. In a large finite number of trials with X results of A and Y results of B, the estimated probabilities are

$$\hat{P}(A) = \frac{X}{X+Y} \qquad \hat{P}(B) = \frac{Y}{X+Y} \qquad (2\text{-}1)$$

since true values for probabilities can be obtained only from an infinite number of trials.

Consider a fixed number of components, N_0, under test. After time t, with $N_s(t)$ surviving and $N_f(t)$ failing $[N_s(t) + N_f(t) = N_0]$, the estimated probability of survival is

$$\hat{P}_{\text{survival}} = R(t) = \frac{N_s(t)}{N_s(t) + N_f(t)} = \frac{N_s(t)}{N_0} \qquad (2\text{-}2)$$

As the test proceeds, $N_s(t)$ gets smaller and therefore $R(t)$ decreases. In a test such as this, without replacement, the reliability R, is truly a function of time. The estimated probability of failure is

$$\hat{P}_{\text{failure}} = Q(t) = \frac{N_f(t)}{N_s(t) + N_f(t)} = \frac{N_f(t)}{N_0} \quad (2\text{-}3)$$

These are complementary, i.e., mutually exclusive, events as it is obvious that $R(t) + Q(t) = 1$.

Equation (2-2) can be rewritten

$$R(t) = \frac{N_s(t)}{N_s(t) + N_f(t)} = \frac{N_0 - N_f(t)}{N_0} = 1 - \frac{N_f(t)}{N_0} \quad (2\text{-}4)$$

Upon differentiating Eq. (2-4)

$$\frac{dR(t)}{dt} = -\frac{1}{N_0}\frac{dN_f(t)}{dt} \quad (2\text{-}5)$$

As dt approaches zero, we obtain the instantaneous probability, i.e., the probability density function (pdf, see Appendix A)

$$\frac{dR(t)}{dt} = -f(t) \quad (2\text{-}6)$$

2-2 INSTANTANEOUS FAILURE RATE

A rearrangement of Eq. (2-5) gives an expression for the rate at which components fail

$$\frac{dN_f(t)}{dt} = -N_0\frac{dR(t)}{dt} \quad (2\text{-}7)$$

$$\frac{dN_s(t)}{dt} = \frac{d(N_0 - N_f(t))}{dt} = -\frac{dN_f(t)}{dt} \quad (2\text{-}8)$$

which is also the negative of the rate at which components survive. This failure rate [Eq. (2-7)] can be interpreted as the number of components failing in time interval dt, between t and $(t + dt)$. This is equivalent to the rate at which the component population still under test at time t is failing.

If both sides of Eq. (2-7) are divided by $N_s(t)$,

$$\frac{1}{N_s(t)}\frac{dN_f(t)}{dt} = -\frac{N_0}{N_s(t)}\frac{dR(t)}{dt} \equiv \lambda(t) \quad (2\text{-}9)$$

where $\lambda(t)$ is defined as the instantaneous failure rate (sometimes known as hazard rate, i.e., the instantaneous probability of failure per component). There is an inherent assumption that $\lambda(t)$ is any variable and integrable function of time t. Substitution of Eq. (2-2) into Eq. (2-9) gives

$$\lambda(t) = - \frac{1}{R(t)} \frac{dR(t)}{dt} \quad (2\text{-}10)$$

which is the most general expression of failure rate.

We know that $dR(t)/dt$ represents the slope of the reliability $[R(t)]$ curve at any time t. This slope is always negative. The largest numerical value occurs at $t = 0$, and it decreases to zero when t is infinite. A plot of $dN_f(t)/dt$ against time gives a distribution, over time, of the failures of all N_0 original components. A plot of $(1/N_0)[dN_f(t)/dt]$ against time, however, gives a distribution, over time, of failures on a per-component basis, i.e., failure frequency per component. This is in effect a unit frequency function, or a failure density function $f(t)$ and, therefore, a restatement of Eq. (2-6). The total area under this latter curve is equal to unity.

Combination of Eqs. (2-10) and (2-6) gives an alternative expression for the instantaneous failure rate

$$\lambda(t) = \frac{f(t)}{R(t)} \quad (2\text{-}11)$$

In other words, the instantaneous failure rate (at time t) equals the failure density function divided by the reliability, with both of the latter evaluated at time t. This is a completely general statement.

2-3 GENERAL RELIABILITY FUNCTION

Rearrangement of Eq. (2-10) gives

$$\lambda(t)\, dt = - \frac{dR(t)}{R(t)}$$

Integration from time 0 to time t gives

$$\int_0^t \lambda(t)\, dt = -\ln R(t) \Big]_1^{R(t)}$$

$$\ln R(t) = - \int_0^t \lambda(t)\, dt$$

Solution for reliability R, knowing that at $t = 0$, $R(t) = 1$ [Eq. (1-2)], gives the general reliability function

$$R(t) = e^{-\int_0^t \lambda(t)\, dt} = \exp\left[-\int_0^t \lambda(t)\, dt\right] \quad (2\text{-}12)$$

This is a mathematical description of reliability in the most general way possible. It is independent of the specific failure distribution involved.

The case of constant failure rate, i.e., $\lambda(t)$ independent of time, is of special interest

$$-\int_0^t \lambda(t)\, dt = -\lambda t$$
$$R(t) = e^{-\lambda t} \quad (2\text{-}12a)$$

This special case will be discussed in much more detail later, i.e., Sec. 3-5.

2-4 GRAPHICAL REPRESENTATION OF RELIABILITY

Rearrangement of Eq. (2-6) gives

$$f(t)\, dt = -dR(t)$$

Integrating

$$\int_0^t f(t)\, dt = -\int_1^{R(t)} dR(t)$$
$$= -[R(t) - 1] = 1 - R(t)$$

But

$$\int_0^t f(t)\, dt + \int_t^\infty f(t)\, dt = \int_0^\infty f(t)\, dt = 1$$

Rearranging

$$\int_0^t f(t)\, dt = 1 - \int_t^\infty f(t)\, dt = 1 - R(t)$$

Therefore, reliability is

$$R(t) = \int_t^\infty f(t)\, dt \quad (2\text{-}13)$$

We know, however, that

$$R(t) + Q(t) = 1$$

Therefore, unreliability is

$$Q(t) = \int_0^t f(t)\, dt \quad (2\text{-}14)$$

These are shown schematically in Fig. 2-1.

In keeping with this concept, let us determine the reliability, i.e., the probability of functioning properly, of a system between times t_1 and t_2 after the start of a "mission" at time $t = 0$. The probability of failure is

$$Q_{t_2-t_1} = \int_{t_1}^{t_2} f(t)\, dt = Q(t)\Big]_{t_1}^{t_2} = Q(t_2) - Q(t_1)$$

The probability of nonfailure in the time interval t_1 to t_2 is

$$R_{t_2-t_1} = 1 - Q_{t_2-t_1} = 1 - [Q(t_2) - Q(t_1)]$$
$$= 1 - R(t_1) + R(t_2) \quad (2\text{-}15)$$

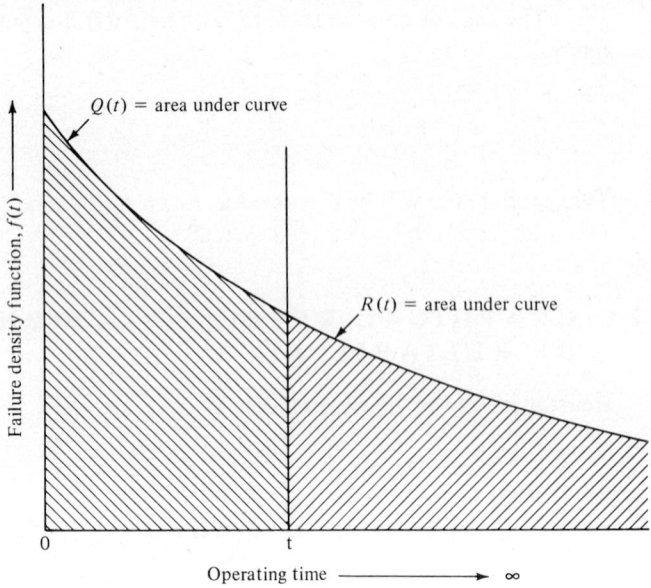

FIGURE 2-1
Schematic representation of relationship between reliability and unreliability.

This answer is not, however, the reliability of the system for proper functioning from t_1 to t_2 because the system may fail before time t_1, i.e., in the period between $t = 0$ and $t = t_1$. $R_{t_2-t_1}$ is merely the *a priori* probability that the system will not fail in the specified time interval. This is *not* identical with the probability that the system will operate in the same time interval, i.e., the system's reliability.

System reliability is the probability that the system will survive until time t_2. This is the same as the probability that it will not fail between $t = 0$ and $t = t_1$ and *also* that it will not fail from $t = t_1$ to $t = t_2$. Therefore, the true reliability is $R(t_2)$.

2-5 MORTALITY CURVE

Consider a population of homogeneous components from which a very large sample is taken and placed in operation at time $T = 0$. (T is age, in contrast with t, commonly used for operational time or mission life.) The population will initially show a high failure rate. This will decrease rather rapidly as indicated in Fig. 2-2. This period of decreasing failure rate is called the "early life period," "burn-in period," "debugging period," "break-in period," "shakedown period," or "infantile mortality period." During this period, failures occur which are due to design or manufacturing weaknesses. In other words,

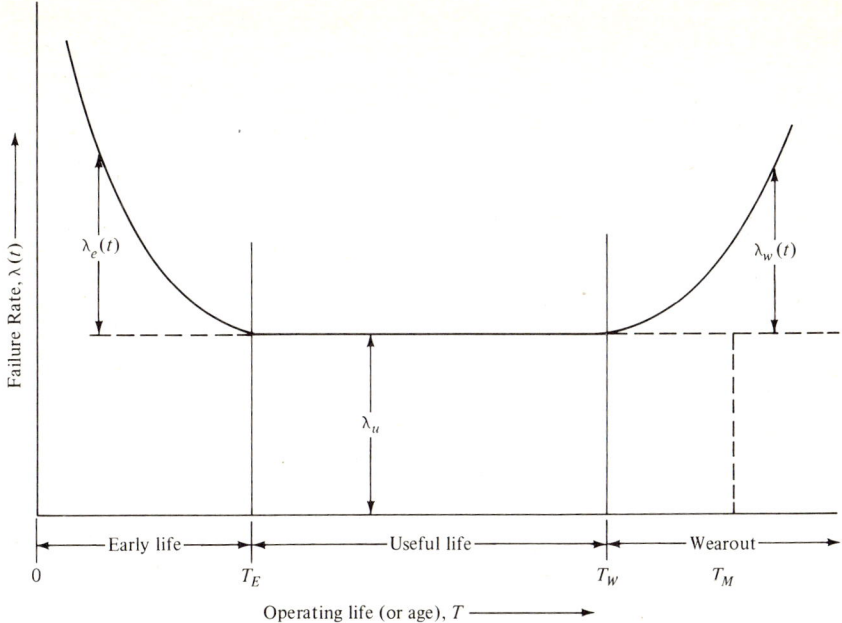

FIGURE 2-2
Mortality curve: Failure rate versus age (schematic).

failure is due to weak or substandard components in which the probability of failure depends on how long the component has been operating.

When the substandard components have all failed at age T_E, the failure rate stabilizes at an approximately constant value. This period of operation is known as useful life since the components can be used to the greatest advantage. Failures during useful life are known as "random," "chance," or "catastrophic" since they occur randomly and unpredictably.

When the components reach age T_W, the failure rate again increases when degradation failures begin to appear as a consequence of aging or wear when the components are nearing their "rated life." Normally, a very small fraction of the population will have failed by T_W. Of the surviving majority, about one-half will fail in the period between T_W and T_M where T_M is the mean wearout life (mean life) of the population.

Reliability, in general, is concerned with all three periods. For moderately reliable equipment, however, early life can be made as short as desired (even eliminated) by proper design, fabrication, and assembly, or by deliberate burn-in periods. Component burn-in before assembly, followed by debugging of the assembly or system is highly desirable. It is an absolute necessity for missile, rocket, and space systems. If design and application are correct, the wearout period should never be reached within operating life. For equipment or systems which must operate satisfactorily over very long or extended periods

14 INTRODUCTION TO RELIABILITY IN DESIGN

of time, the incidence of wearout can be postponed almost indefinitely. Components are replaced as they fail during useful life, and each component is replaced (even if it has not failed or has any indication of imminent failure) not later than at the end of its useful life.

2-6 EXPANDED RELIABILITY FUNCTION

In Eq. (2-12), the general reliability function was

$$R(t) = \exp\left[-\int_0^t \lambda(t)\,dt\right] \qquad (2\text{-}12)$$

We can now expand this in terms of a more detailed failure rate as shown in Fig. 2-2. For early life

$$\lambda_e(t) = \lambda_e(\tau) + \lambda_u + \lambda_w(\tau)$$

but $\lambda_w(\tau)$ is negligible during early life. For useful life

$$\lambda_u(t) = \lambda_e(\tau) + \lambda_u + \lambda_w(\tau)$$

but both $\lambda_e(\tau)$ and $\lambda_w(\tau)$ are negligible during useful life.
For wearout

$$\lambda_w(t) = \lambda_e(\tau) + \lambda_u + \lambda_w(\tau)$$

but $\lambda_e(\tau)$ is negligible during wearout.

A population consisting of several subpopulations can have early life, useful life, and wearout simultaneously. In this situation, the general reliability function becomes

$$R(t) = R_e(t)R_u(t)R_w(t) \qquad (2\text{-}16)$$

which is equivalent to

$$R(t) = \exp\left\{-\int_0^t [\lambda_e(\tau) + \lambda_u + \lambda_w(\tau)]\right\} d\tau \qquad (2\text{-}17)$$

This can be rewritten as

$$R(t) = \left\{\exp\left[-\int_0^t \lambda_e(\tau)\,d\tau\right]\right\} (e^{-\lambda_u t}) \left\{\exp\left[-\int_0^t \lambda_w(\tau)\,d\tau\right]\right\} \qquad (2\text{-}18)$$

PROBLEMS

2-1 List all types, causes, or sources of failure you can think of and classify them as:
  (*a*) early failures
  (*b*) random failures
  (*c*) wearout failures

2-2 Demonstrate the validity of
  (*a*) Eqs. (2-7) and (2-8)
  (*b*) Eq. (2-11)
  (*c*) Eq. (2-12)
  (*d*) Eq. (2-13)
  (*e*) Eq. (2-15)
  (*f*) Eq. (2-16)
  (*g*) Eq. (2-17)
  (*h*) Eq. (2-18)

3
COMPONENT LIFE

3-1 COMPONENT MORTALITY

Consider Eq. (2-6), which we now restate in somewhat different form, with a broader interpretation than before, i.e.,

$$f(t) = P[t < T \leq (t + dt)] = -\frac{dR(t)}{dt} \quad (2\text{-}6)$$

We also know the general reliability function is

$$R(t) = \exp\left[-\int_0^t \lambda(\tau)\, d\tau\right] \quad (2\text{-}12)$$

Upon differentiating,

$$\frac{dR(t)}{dt} = -\lambda(t) \exp\left[-\int_0^t \lambda(\tau)\, d\tau\right]$$

which, upon substitution in Eq. (2-6), gives

$$f(t) = \lambda(t) \exp\left[-\int_0^t \lambda(\tau)\, d\tau\right] \equiv m(t) \quad (3\text{-}1)$$

in which $m(t)$ is known as the *mortality* of the component. The instantaneous failure rate then can be written either as Eq. (2-11) or as

$$\lambda(t) = \frac{m(t)}{R(t)} \qquad (3\text{-}2)$$

It is obvious that $m(t)$ and $\lambda(t)$ are two entirely different functions in mathematical terms and in concept.

The mortality $m(t)$ is the *unconditional probability* that the component will fail in the time interval dt between t and $t + dt$. The instantaneous failure rate $\lambda(t)$ is the *conditional probability* that the component will fail in the same time interval, given it has reached age T without failure. By analogy, the probability that a newborn child will die at an age between 108 and 109 years [corresponding to $m(t)\, dt$] is obviously very small. The probability of dying in that same period, provided the person has lived to the age of 108 [corresponding to $\lambda(t)\, dt$] is much greater.

3-2 MEAN TIME TO FAILURE (MTTF)

The mean time to failure (MTTF), specifically the mean time to first failure, can be found by applying Eq. (A-17).

$$\bar{T} = \text{MTTF} = \int_0^\infty t\, m(t)\, dt = -\int_0^\infty t\, \frac{dR(t)}{dt}\, dt$$

Integrating by parts,

$$\bar{T} = \text{MTTF} = -t\, R(t)\Big|_0^\infty + \int_0^\infty R(t)\, dt$$

From Eq. (1-2), $R(0) = 1$ and $R(\infty) = 0$. Thus, the first term equals zero and MTTF becomes

$$\bar{T} = \text{MTTF} = \int_0^\infty R(t)\, dt = \int_0^\infty \left\{ \exp\left[-\int_0^t \lambda(\tau)\, d\tau\right]\right\} dt \qquad (3\text{-}3)$$

For the special case of useful life with a constant failure rate,

$$R(t) = e^{-\lambda t}$$

$$\text{MTTF} = \int_0^\infty e^{-\lambda t}\, dt = \frac{1}{\lambda} \qquad (3\text{-}3a)$$

3-3 MEAN TIME BETWEEN FAILURES (MTBF)

Strictly speaking, "mean time to failure" should be used in the case of simple components which are not repaired when they fail but are replaced by good components. Likewise, "mean time between failures" should be used with repairable equipment or systems. It has become customary, however, to use MTBF for both nonrepairable components and repairable equipment and systems. In any event, it represents the same statistical concept of the mean time at which failures occur. This mean must be known in order to make the probability calculations which are necessary for evaluation of reliability of components and systems.

As indicated, MTTF is the mean time to first failure. Conversely, MTBF is the mean time between two successive component failures. These are not necessarily failures of "identical" components and generally will not be. The MTBF is normally conceived as being the mean time between the nth and the $(n+1)$th failure in a system when n is relatively large. The relationship between the MTBF of a system and the MTTFs of the individual components is

$$\frac{1}{\text{MTBF}} = \sum_{j=1}^{m} \frac{1}{\bar{T}_j} \qquad (3\text{-}4)$$

where the system has m components, all of different ages, each of which is replaced immediately on failure, and $\bar{T}_j$ is the MTTF of the jth component.

It is obvious that MTBF will be a function of time. When a system is first operated with all new components, the MTTF and the MTBF are identical. After that, the MTBF will fluctuate until, after many failures and replacements, it will stabilize at the value given by Eq. (3-4). For the special case of useful life, Fig. 2-2, MTTF and MTBF are equal, a fact which contributes to confusion between the two concepts.

3-4 A PRIORI—A POSTERIORI

An "a priori probability" is one calculated prior to knowing any information which may be obtained from an experiment. Thus, it is "before the facts," i.e., an estimate which is concerned only with an event at time t. It is therefore, in most cases, only a partial answer.

An "a posteriori probability" is one calculated after the results of an experiment are known. Thus, it is "after the facts," or a full and complete answer because it considers what has preceded the event as well as the event itself at time t.

Consider the question, "What is the probability of a device's failure or survival in a given time interval, if the device survives until the beginning of that time interval?" This introduces a condition, "if it survives," and thereby creates a case of conditional probability. From the theorem on conditional probabilities [Eq. (A-5)], the probability of survival in the time interval $(t_2 - t_1)$, given that the device has survived until time t_1, is the a priori probability of survival in the time interval $(t_2 - t_1)$ divided by the cumulative probability of surviving from $t = 0$ to $t = t_1$. The resulting probability (reliability) is the a posteriori probability of survival in the time interval $(t_2 - t_1)$.

3-5 USEFUL LIFE

The useful life period (Fig. 2-2) is one in which there is a constant failure rate. In this case, Eq. (2-12a) applies,

$$R(t) = \exp\left(-\int_0^t \lambda \, dt\right) = e^{-\lambda t} \qquad (3\text{-}5)$$

indicating that the mortality, Eq. (3-1),

$$m(t) = f(t) = -\frac{d(e^{-\lambda t})}{dt} = \lambda e^{-\lambda t} \qquad (3\text{-}6)$$

has a negative exponential distribution.

As will be seen, the calculation for useful life reliability is much simpler than for early life or wearout. This is fortuitous, since most practical interest centers on the useful life period. In addition, as indicated in Sec. 2-5, early life and wearout can be minimized by appropriate action.

The reliability, i.e., probability of survival from $t = 0$ to time t is

$$R(t) = e^{-\lambda t} = e^{-t/\text{MTBF}} \qquad (3\text{-}5)$$

The unreliability, i.e., probability of failure in the same time interval is

$$Q(t) = 1 - R(t) = 1 - e^{-\lambda t} \qquad (3\text{-}7)$$

The probability of survival from $t = t_1$ to $t = t_2$ is

$$R(t_2 - t_1) = e^{-\lambda(t_2 - t_1)}$$

All three of these are a priori probabilities.

What is the unreliability, i.e., probability of failure, of a device during a period of t h when the device has already operated for an age of T h? (This is an a posteriori probability.)

$$\begin{aligned} P_{\text{failure}} &= \int_T^{T+t} m(T) \, dT = \int_T^{T+t} \lambda e^{-\lambda T} \, dT \\ &= -\lambda/\lambda [e^{-\lambda T}]_T^{T+t} = e^{-\lambda T} - e^{-(T+t)} \\ &= e^{-\lambda T}(1 - e^{-\lambda t}) \end{aligned} \qquad (3\text{-}8)$$

This expression [Eq. (3-8)] is an a priori probability of failure in the period from T to $T + t$. To obtain the a posteriori probability of failure, Eq. (3-8) must be divided by the probability of survival up to age T, given by Eq. (3-5). Therefore, the unreliability is

$$\frac{e^{-\lambda T}(1 - e^{-\lambda t})}{e^{-\lambda T}} = 1 - e^{-\lambda t} \qquad (3\text{-}9)$$

This expression is identical with Eq. (3-7), i.e., the cumulative probability of failure from $T = 0$ to $T = t$. We conclude that a priori and a posteriori probabilities are identical in the case of useful life, i.e., a negative exponential distribution for mortality in which there is a constant failure rate. While this is true for useful life, it is *not* true for other mortality distributions and, therefore, the operation of division performed in obtaining Eq. (3-9) is necessary in general.

EXAMPLE 3-1   (*a*) If a device has a failure rate of $0.5 \times 1\$^{-5}$ failures/h = 5×10^{-6} failures/h = 0.5%/1000 h, what is its reliability for an operating period of 100 h?   If there are 10,000 items in the test, how many failures are expected in 100 h?

$R(t) = R(100) = e^{-0.5 \times 10^{-5} \times 10^2} = e^{-0.0005}$
$R(100) \simeq 1 - 0.0005 = 0.9995$ (see Prob. 3-1)
$N_s = N_0 R(t) = 10,000 \times 0.9995 = 9995$
$N_f = N_0 - N_s = 5$ failures expected

(*b*)   From Eq. (3-3*a*), MTTF and MTBF are equal to each other and to $1/\lambda$. For the device above, MTBF is 200,000 h.   What is the reliability for an operating time equal to the MTBF?

For operating time, t = MTBF

$$R(t) = e^{-t/t} = e^{-1} = 0.3679$$

(*c*)   If the useful life is 100,000 h, what is the reliability for operating over its useful life?

$$R(t) = e^{-0.5 \times 10^{-5} \times 10^5} = e^{-0.5} = 0.6065$$

(*d*)   What is the probability for its surviving the first 100 h?   If the device has survived for 99,900 h, what is the probability for its surviving the next 100 h?

In both cases, $R(t) = 0.9995$, which is the reliability for any 100-h operating period during useful life.

COMMENT: For an operating time equal to the MTBF there is only about a 37 percent probability of failureless operation.   For one-tenth of the MTBF, the reliability is approximately 90 percent, while it is about 95 percent for MTBF/20 and 99 percent for MTBF/100.   ////

Within useful life, the reliability of a device is the same for operating periods of equal length.   In effect, the device is always as good as new, since the failure rate is constant.   Wearout has not yet had an opportunity to cause any degradation which makes the device more susceptible to failure.   The probability that the device will have a random failure remains the same for periods of equal length throughout the entire useful life.   During useful life, the failure rate is not affected by time or aging.   An automobile or similar device with exponential reliability (useful life), after 6 months or a year, is, therefore, "as good as new" (which, incidentally, need not be very good nor very reliable).   There is no reason, however, why the MTBF (or MTTF) should change after some use in the useful life range.

The MTBF (obtained in useful life) is often much larger than the mean wearout life T_M. If the failure rate in useful life is very small, the MTBF can reach several thousand or even millions of hours.   However, if a device has an MTBF of 10^6 h, it does not mean that the device can be used for 10^6 h. As indicated in Example 3-1, the device would have only a 36.8 percent chance of successful operation for its MTBF, provided it is still within the useful life period. Since life is limited by wearout, the probability of surviving for a period equal to the MTBF is very remote.

If this device has a mean life T_M of 10,000 h, the life T_W might well be about 6000 h. If we wish to avoid failure during operation, we can only use the device for about 6000 h. The low failure rate which gives an MTBF of 10^6 h is effective only during the first 6000 h. If the device does not fail during that 6000-h period, it must be replaced if we wish to continue operation at the high level of reliability indicated by the MTBF of 10^6 h.

The MTBF simply indicates how reliable a component is within its useful life. It has no relationship to wearout life. Components with large MTBFs will give highly reliable systems, provided components are not operated beyond T_W. We must remember, however, that *some* chance failures will occur even in very highly reliable systems.

The Weibull distribution [Eq. (A-37)] can be used for all three stages of the age of a component or a device. The value of the Weibull slope b is always positive. For useful life, $b = 1$. If x_0 is zero, this becomes a special case where

$$f(t) = (1/\theta)e^{-t/\theta} \quad (3\text{-}6a)$$

It is seen that this is the same as Eq. (3-6), if $1/\theta = \lambda$, where θ, the characteristic life, is equivalent to the MTBF. This is often done for convenience.

3-6 WEAROUT

A complete reliability program cannot neglect wearout studies by concentrating only on random failure studies. Wearout studies are absolutely necessary to establish preventive maintenance policies (for long-life devices) or to assure that wearout phenomena cannot affect a system during a mission, especially a one-shot situation, such as a space probe or an Apollo mission.

Once degradation or deterioration of a component or device starts (at about T_W), the failure rate begins to increase, and wearout becomes superposed on the constant failure rate of the exponential failure frequency. These failures tend to cluster around the mean wearout life T_M. The instantaneous failure rate is given by Eq. (2-11) or (3-2). Experience shows that wearout usually follows the normal (or lognormal) distribution, i.e.,

$$m(T) = f(T) = \frac{1}{\sigma\sqrt{2\pi}} e^{-(T-T_M)^2/2\sigma^2} \quad (3\text{-}10)$$

where T is component age (accumulated operating time since new), T_M is mean wearout life, and $\sigma = \sqrt{[\Sigma(T - T_M)]^2/(N - 1)}$. The probability density function (or mortality), i.e., the two parameters T_M and σ, must be evaluated from actual test data, by direct calculation, by using histograms, or by other methods. From Eq. (2-13), we know that the reliability at age T is

$$R(T) = \int_T^\infty f(T)\, dT \qquad (3\text{-}11)$$

Computation is sometimes made easier by using the standardized function

$$\phi(z) = (1/\sqrt{2\pi})e^{-z^2/2} \qquad (3\text{-}12)$$

where

$$z = \frac{T - T_M}{\sigma}$$

Thus

$$f(T) = \frac{\phi(z)}{\sigma}$$

and

$$\lambda(T) = \frac{f(T)}{R(T)} = \frac{\phi(z)}{\sigma R(T)}$$

giving reliability

$$R(T) = \int_T^\infty f(T)\, dT = \int_z^\infty \phi(z)\, dz = R(z) \qquad (3\text{-}11a)$$

If we define

$$\lambda(z) \equiv \sigma \lambda(T)$$

then we have a standardized failure rate

$$\lambda(z) = \frac{\phi(z)}{R(z)} \qquad (3\text{-}11b)$$

This standardized failure rate (in failures per standard deviation) can be calculated. Upon multiplying this number by the standard deviation (in hours), we obtain the actual failure rate (in failures per hour).

It should be noted that the normal probability distribution function must be extended from negative infinity to positive infinity for the area under the curve to be unity. This is not possible for time-dependent events, e.g., a new component enters service at $T = 0$ and not at negative infinity. The normal distribution is a good approximation, however, as can be seen from Table 3-1. It is obvious that the value of the integral which is neglected up to the start of the wearout period is very small. When there are thousands (or more) of components in a system, however, the wearout probabilities of the individual components add up and the probability of the system surviving wearout decreases rapidly. The choice of replacement time (T_W, i.e., the number of standard deviations before mean wearout time, T_M) must be made very carefully.

A major concern is determining the reliability for an operating time t given an age T, i.e., determining $R_w(T,t)$. We know that any area under the curve, from T_1 to T_2, i.e., $\int_{T_1}^{T_2} f(T)\, dT$, represents the percentage of original N components which fail in the interval from T_1 to T_2. This also represents the a priori probability of any individual component of the original population failing in that time interval when it is put into service at time $T = 0$ when it is new.

The a priori probability of failure, assuming a normal distribution, in the interval $T_M - 3\sigma$ to $T_M - 2\sigma$ is 2.14 percent. There is the same probability of failure in the interval $T_M + 2\sigma$ to $T_M + 3\sigma$. Judged at time $T = 0$, i.e., when it is new, the component has the same probability of failing in the interval $T_M + 2\sigma$ to $T_M + 3\sigma$ as it does of failing in the interval $T_M - 3\sigma$ to $T_M - 2\sigma$. That is, the device has the same chance of being better than average that it does of being worse than average. These equal probabilities, however, do *not* have the same meaning in terms of reliability, i.e., from the viewpoint of the item surviving.

In the first case, survival from $T = 0$ to $T = T_M - 3\sigma$ has a probability of 99.865 percent. In the second case, the probability of survival from $T = 0$ to $T = T_M + 2\sigma$ is only 2.275 percent. The 2.14 percent probability of failure in the two equal time intervals must be weighed against the probability of survival up to the beginning of the respective time interval. The statement that the component has a probability of failure of only 2.14 percent in the interval from $T_M + 2\sigma$ to $T_M + 3\sigma$ has no meaning for reliability unless we also know that its probability of failure before $T_M + 2\sigma$ is 97.725 percent. Thus, it is doubtful that the component will ever reach age $T_M + 2\sigma$. The 2.14 percent is an a priori probability of failure which must be combined with the probability of failure from $T = 0$ to $T_M + 2\sigma$, the beginning of the time interval. In this second case, the probability of failure up to age $T_M + 3\sigma$ is 99.865 percent. The probability of survival (i.e., reliability) beyond $T_M + 3\sigma$ is only 0.135 percent.

The probability of failure in the time interval T_1 to T_2, given that the component has survived up to age T_1, is the a priori probability of failure in the interval $T_2 - T_1$, divided by the cumulative probability of surviving from $T = 0$ to T_1,

$$F_{T_2 - T_1} = \frac{\int_{T_1}^{T_2} f(T)\, dT}{R(T)}$$

In a like manner, the probability of survival from age T for mission time t, given that the component has survived up to age T, is the a priori probability

Table 3-1 INTEGRAL OF NORMAL DISTRIBUTION VALUE TO START OF WEAROUT PERIOD

Multiple of standard deviation, n	$T_W = T_M - n\sigma$	$\int_{-\infty}^{T} f(T)\, dT$
3	$T_M - 3\sigma$	0.00135
4	$T_M - 4\sigma$	0.0000317
4.5	$T_M - 4.5\sigma$	0.0000034
5	$T_M - 5\sigma$	0.000000287

of survival for the mission time divided by the cumulative probability of surviving from $T = 0$ to age T:

$$R_w(T,t) = \frac{R(T+t)}{R(T)}$$

$$R_w(T,t) = \frac{\int_{T+t}^{\infty} f(T)\,dT}{\int_{T}^{\infty} f(T)\,dT} \qquad (3\text{-}13)$$

where $f(T)$ for wearout is assumed to be a normal distribution.

EXAMPLE 3-2   Assuming wearout of a given subsystem is normally distributed with a mean wearout time T_M of 10,000 h and a standard deviation of 1000 h, determine the reliability for an operating time of 400 h if (a) the age of the component is 9000 h, (b) the age of the component is 11,000 h.

Using Eq. (3-13),

(a) $R_w(T,t) = \dfrac{\int_{T_M - 0.6\sigma}^{\infty} f(T)\,dT}{\int_{T_M - \sigma}^{\infty} f(T)\,dT} = \dfrac{0.7251}{0.8413} = 0.863$

(b) $R_w(T,t) = \dfrac{\int_{T_M + 1.4\sigma}^{\infty} f(T)\,dT}{\int_{T_M + \sigma}^{\infty} f(T)\,dT} = \dfrac{0.0818}{0.1587} = 0.516$ ////

The Weibull distribution—with b, the Weibull slope, greater than 1—is sometimes used instead of the normal distribution. In this case, Eq. (3-13) becomes

$$R_w(T,t) = \frac{\exp\left[-\left(\dfrac{T + t - T_w}{\theta - T_w}\right)^b\right]}{\exp\left[-\left(\dfrac{T - T_w}{\theta - T_w}\right)^b\right]}$$

$$R_w(T,t) = \exp\left\{-\left[\left(\frac{T + t - T_w}{\theta - T_w}\right)^b - \left(\frac{T - T_w}{\theta - T_w}\right)^b\right]\right\} \qquad (3\text{-}14)$$

3-7 EARLY LIFE

Early failures can adversely affect system reliability as can random and wearout failures. The probabilistic treatment of early failures is less important than that of the other two, however, since early failures can be eliminated by means of systematic burn-in or debugging procedures. Early failures should not affect reliability when debugging is complete and the system is put into operation.

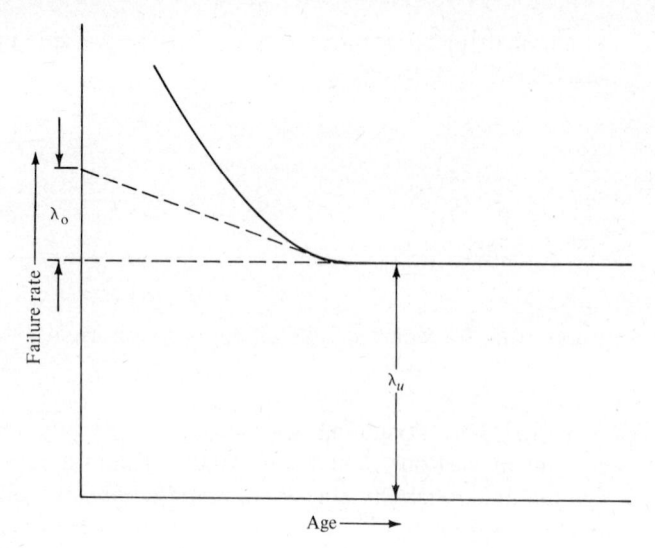

FIGURE 3-1
Segment of Fig. 2-2, showing method of estimating the "constant" in Eq. (3-15).

The time and effort required to eliminate early failures is a matter of experience. For the sake of completeness, however, a brief discussion of early failure reliability is in order.

The Weibull distribution is often used with the slope b between 0 and 1. Equation (3-14) applies with the different value of the slope and the replacement of T_w by 0. One problem arises in that the instantaneous failure rate becomes infinite at $T = 0$.

Since the early failure rate is an exponential form, the following approximation is sometimes used:

$$\lambda_e(T) = \lambda_0 e^{-\alpha T} \qquad (3\text{-}15)$$

where $\alpha = 1/E(t)$
$E(t)$ is known as the mean debugging time and is

$$E(t) = m_e \left(1 + \frac{1}{2} + \frac{1}{3} + \cdots + \frac{1}{N_e}\right) \qquad (3\text{-}16)$$

where m_e is the expected time to failure of a substandard component and there are N_e substandard components. The constant λ_0 is determined as indicated in Fig. 3-1. For high system reliability in operation, the actual debugging time should extend to at least $5E(t)$. The exponent α is generally found from fitting a curve to the available data. This is not a very satisfactory approach, but no better scheme appears available.

Reliability becomes

$$R_e(T) = \exp\left[-\int_0^T \lambda_e(T)\,dT\right]$$

$$R_e(T) = \exp\left[-(\lambda_0/\alpha)(1 - e^{-\alpha T})\right] \quad (3\text{-}17)$$

The probability of survival from age T for mission time t, given that the component has survived up to age T, is

$$R_e(T,t) = \frac{\exp\left[-(\lambda_0/\alpha)(1 - e^{-\alpha(T+t)})\right]}{\exp\left[-(\lambda_0/\alpha)(1 - e^{-\alpha T})\right]}$$

$$R_e(T,t) = \exp\left\{-(\lambda_0/\alpha)[e^{-\alpha T}(1 - e^{-\alpha t})]\right\} \quad (3\text{-}18)$$

3-8 EXPANDED RELIABILITY FUNCTION

We are now in a position to summarize the general reliability function. In Fig. 2-2, we saw the mortality curve in which the failure rate is shown as a function of age. We can expand this by adding mortality and reliability as shown in Fig. 3-2.

Equation (2-17) is an expansion of Eq. (2-12), the general reliability function. This equation can be further expanded to express the probability of survival of a device from age T for mission time t, given that the device has sur-

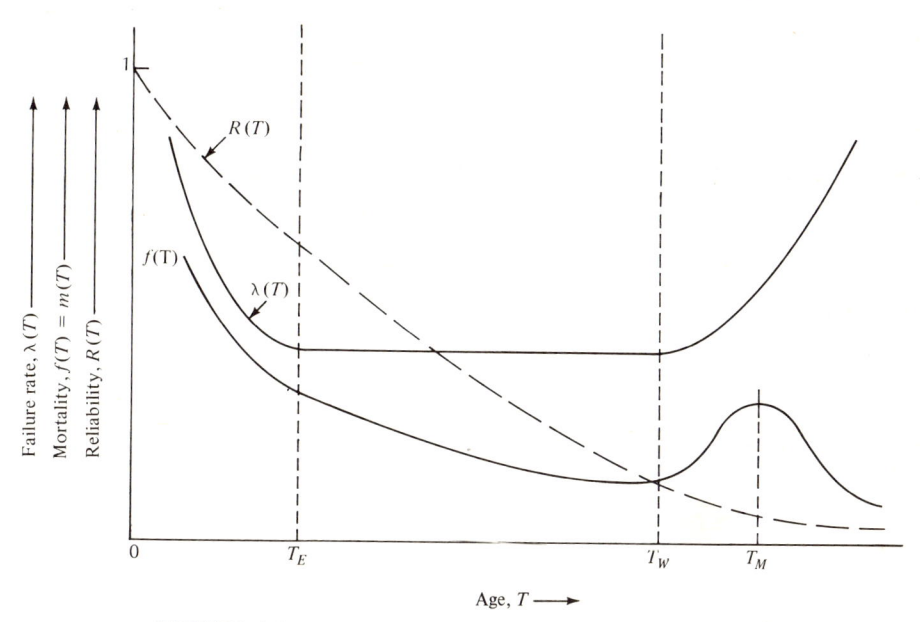

FIGURE 3-2
Schematic mortality curve showing failure rate, mortality, and reliability.

vived up to age T, i.e.,

$$R_{e,u,w}(T,t) = \left(\exp\left\{-\frac{\lambda_0}{\alpha}\left[e^{-\alpha T}(1 - e^{-\alpha t})\right]\right\}\right)$$

$$e^{-\lambda t}\left\{\frac{\int_{T+t}^{\infty} \exp\left[-\frac{1}{2}\left(\frac{T - T_M}{\sigma}\right)^2 dT\right]}{\int_{T}^{\infty} \exp -\frac{1}{2}\left(\frac{T - T_M}{\sigma}\right)^2 dT}\right\} \quad (3\text{-}19)$$

where the terms relate to early life [based on the approximation for $\lambda_e(T)$ although the Weibull distribution can be used], useful life, and wearout life (based on the normal distribution although the Weibull can be used), reading from left to right. If $T \geq 5/\alpha$, then R_e is essentially equal to unity. If $T + t \leq (T_M - n\sigma)$ (see Table 3-1), then R_w is essentially equal to unity.

So far, we have discussed basic (mathematical) theory, and we have developed general expressions. The discussion has been in terms of components or devices. There has been an implicit assumption of application to "identical" items in a population, all items being "loaded" and operated in the same manner. There is no necessity for such an assumption. The theory and equations apply equally well (although complexity increases) as we go from parts through components, units, devices, subassemblies, assemblies, subsystems, and systems to "super-systems" of great complexity.

Once correct values for failure rates or reliabilities (or good estimates of these values) of components in systems are available, we can perform exact calculations of system reliabilities even when the systems are the most complex combinations of components conceivable. The exactness of the results does not depend on the probability calculations (these are quite exact) but on the exactness of the data on which the calculations are based.

Reliability calculations are a necessary and integral part of the design of a system. Once a device or system has been designed and production started, however, relatively little can be done about its reliability.

PROBLEMS

3-1 In Example 3-1, we find that reliability was approximated by subtracting the exponent λt from unity. This is a good approximation for small values of λt. Determine the limits of λt for which this is valid for 4, 5, and 6 decimal places.

3-2 Demonstrate the validity of (a) Eq. (3-1), (b) Eq. (3-11a).

3-3 What is the reliability of an engine with a failure rate of $10/10^7$ h for an operating time of 100 h? 1000 h? 10,000 h?

3-4 The mortality of a given component is $0.00002\, e^{-\lambda t}$. What is the failure rate, the MTTF, the MTBF, and the reliability for an operating period of 0.1 MTBF and 0.2 MTBF?

3-5 Determine the number of replacement engines required on an annual basis to keep 2000 engines in operation. Two hundred of these operate from 0 to 200 h of age; each of 1600 operates for 7500 h of useful life during the year; and the remaining 200 operate from 50,000 to 50,200 h of age. Company policy dictates that no engine be operated in the wearout period.

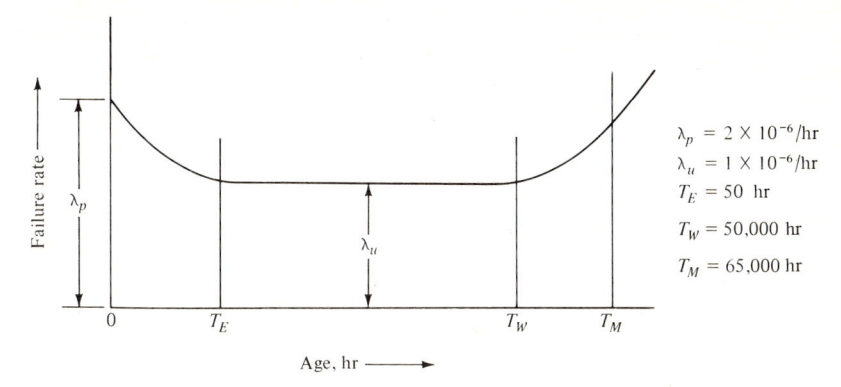

$\lambda_p = 2 \times 10^{-6}/\text{hr}$
$\lambda_u = 1 \times 10^{-6}/\text{hr}$
$T_E = 50$ hr
$T_W = 50,000$ hr
$T_M = 65,000$ hr

3-6 A given component has an MTBF of 10^6 h.
  (a) What is the reliability for an operating period of 10 h?
  (b) What is the reliability for an operating period of 10 h for 5 units in series? For 10 in series?
  (c) What is the reliability for a single component for an operating period of 100 h starting from an age of 1000 h? From an age of 2000 h? What inherent assumption did you make?
  (d) What is the reliability for a component which starts operating at an age of 2100 h in useful life? What assumption did you make?
  (e) Draw conclusions from comparisons of the above calculations.

3-7 (a) Estimate the reliability and failure rate of a test sample of 200 items from a component population if 8 fail during the first hour.
  (b) If failing items are replaced, estimate reliabilities and failure rates when two more fail during the second hour, five more fail during the third hour, four more fail during the fourth and fifth hours, and eight fail during the sixth through tenth hour.
  (c) Estimate reliabilities and failure rates for the data in part (b) if none of the failed items are replaced.
  (d) Compare the results and draw conclusions. What difference, if any, would there be in the results if these were early or useful life tests?

3-8 An engine shaft has a failure rate of 0.5×10^{-7}/h. The seals used with the shaft have a failure rate of 2.5×10^{-7}/h. If a given company has 5000 engines with these shafts and seals and each engine operates 350 days per year in useful life, estimate the number of shafts and seals that must be replaced annually.

3-9 A sample of 150 components is subjected to testing (presumably in useful life). Three failures are found at the end of 400 h; four more at the end of 800 h; two more at the end of 1200 h; four more at the end of 1800 h; and no further failures are found when the test is terminated at 2500 h.
  (a) Estimate the MTBF if failed components are replaced when found.
  (b) Estimate the MTBF if no replacements are made.
  (c) What is the most conservative estimate you could make using these data?

3-10 Two types of components with identical electrical characteristics have different failure rates: The failure rate of component A is $\lambda_A(t) = \text{constant} = 1\%/1000$ h; for B it is $\lambda_B(t) = 10^{-6}t$ where t is in hours. Which of the two components is more reliable for a run of 10, 100, 1000, 10,000 h?

3-11 The failure rate for a certain type of component is $\lambda(t) = \lambda_0 t$ where $\lambda_0 > 0$ and is constant. Find its reliability, mortality, and MTTF. Repeat for $\lambda(t) = \lambda_0 t^{1/2}$.

3-12 The failure rate for a certain type of component is $\lambda(t) = a + bt$ where $a > 0$ and $b > 0$ are constant. Find reliability, mortality, and MTTF.

3-13 A given item has a random failure rate of 10^{-5} failure/h. Wearout is normally distributed with a mean (T_M) of 1500 h and a standard deviation of 150 h.

(a) What is the a priori probability of survival of the item from an age of 1350 h to 1650 h?
(b) What is the a posteriori probability of survival (i.e., the reliability) of the item for the same period?
(c) What is the reliability of the item for an operating period of 15 h starting at an age of 1350 h?

3-14 A small generator has a random failure rate of 10^{-4} failure/h. Assume wearout is normally distributed with a mean of 12,000 h and a standard deviation of 2000 h. What is the reliability for an operating period of 100 h if its age is 11,900 h?

3-15 A component has a random failure rate of 10^{-5} failure/h. $\alpha = \frac{1}{25}$; $\lambda_e = 10^{-4}$ failure/h; $T_M = 1200$ h; $\sigma = 100$ h.
(a) Estimate $R_{e,u,w}(T,t)$ for an operating period of 20 h starting at age 40 h.
(b) Estimate $R_{e,u,w}(T,t)$ for an operating period of 700 h starting at age 600 h.

3-16 A given system has a random failure rate of 10^{-3} %/h, a T_M of 15,000 h, and a standard deviation of 600 h.
(a) An operating period starts at an age of 12,600 h. Determine and plot: $R_u(t)$, $R_w(t)$, and system reliability for an operating period of 900 h.
(b) After completion of part (a), two additional successive periods of 900 h each are carried out. Determine and plot: $R_u(t)$, $R_w(t)$, and system reliability.
(c) What assumption did you make? Would you get different results if you had made different assumptions? What different assumptions are possible? Discuss the ramifications of these differences.

3-17 Suppose that for a given product the probability of an initial failure (failure prior to time $t = \alpha$, i.e., early life) is θ_1, the probability of a wearout failure (failure beyond time $t = \beta$) is θ_2, and that for the interval $\alpha \leq t \leq \beta$ mortality is given by

$$m(t) = \frac{1 - \theta_1 - \theta_2}{\beta - \alpha}$$

(a) Find an expression for the cumulative density function for the interval $\alpha \leq t \leq \beta$.
(b) Find an expression for the reliability for the same interval.
(c) Find an expression for the instantaneous failure rate for the same interval.
(d) Assume an automobile tire is an early failure if it fails during the first 500 mi and a wearout failure if it occurs after 10,000 mi. If the foregoing model holds, and $\theta_1 = 0.03$ and $\theta_2 = 0.85$, sketch the graph of the failure rate function from $t = 500$ to $t = 10,000$ mi.

4
SERIES AND PARALLEL SYSTEMS

4-1 INTRODUCTION

Many systems consist of various components which can be considered to be in series, in parallel, or in combinations of series and parallel. A series system is one in which all components are so interrelated that the entire system will fail if any one of its components fails; a parallel system is one which will fail only if all its components fail.

Consider a switch which can operate properly and also has possible malfunctions such as (1) not closing when required to close; (2) closing before required to close; and (3) once closed, staying closed when required to open. By previous definition, reliability R is the probability that the switch will function properly, e.g., close when it is required to close. Likewise, unreliability Q is the probability that the switch will not function properly. We also know that $R + Q = 1$.

Consider a situation in which we use two independent identical switches X and Y as shown in Fig. 4-1. The criterion of adequate performance for the system is closing the circuit upon command and allowing current to flow. Assume the reliability of a single switch is 0.995. The unreliability is then 0.005.

FIGURE 4-1
Two arrangements of switches to establish flow of current: (a) series, (b) parallel.

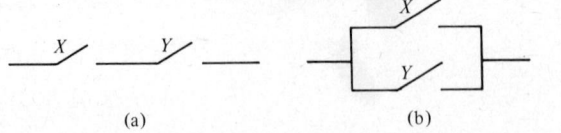

For the case of the two switches in series (Fig. 4-1a) both must close on command to meet the criterion of adequate performance. The probability of this, i.e., the reliability [using Eq. (A-5a)] is

$$P = R = R_X R_Y = (0.995)^2 = 0.990025$$

For the case of the two switches in parallel, there are three ways the criterion of adequate performance can be met:

1. X is closed and Y is open.
2. Y is closed and X is open.
3. Both X and Y are closed.

For case 1:

$$P(X)F(Y) = R_X(1 - R_Y)$$

For case 2:

$$P(Y)F(X) = R_Y(1 - R_X)$$

For case 3:

$$P(X)P(Y) = R_X R_Y$$

All three cases are mutually exclusive. If any one of the three occurs, the circuit is closed. The probability of this, i.e., the reliability [using Eq. (A-4a)] is

$$P(X \text{ or } Y \text{ or both}) = R_X(1 - R_Y) + R_Y(1 - R_X) + R_X R_Y$$

Multiplying and collecting terms

$$P = R = R_X + R_Y - R_X R_Y$$
$$= 2(0.995) - (0.995)^2 = 0.999975$$

As an alternative solution for the parallel system, there is only one way for it to fail, i.e., for both switches to fail to close:

$$F(X)F(Y) = (1 - R_X)(1 - R_Y)$$

But

$$R = 1 - F(X)F(Y)$$
$$R = 1 - (1 - R_X)(1 - R_Y)$$
$$R = R_X + R_Y - R_X R_Y$$

Obviously, this is identical to the previous solution.

FIGURE 4-2
Schematic system with units in series.

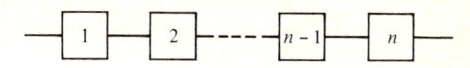

4-2 SERIES SYSTEMS

Assuming n components in Fig. 4-2 are independent and the system will survive for operating time t [with $R(0) = 1$] if, and only if, each one of the components survives to time t, then the system reliability is

$$R_{ss}(t) = R_1(t)R_2(t) \cdots R_n(t) = \prod_{i=1}^{n} R_i(t) \qquad (4\text{-}1)$$

From Eq. (2-12)

$$R_i(t) = \exp\left[-\int_0^t \lambda_i(\tau)\,d\tau\right]$$

$$R_{ss}(t) = \exp\left[-\sum_{i=1}^{n}\int_0^t \lambda_i(\tau)\,d\tau\right] = \exp\left\{-\int_0^t\left[\sum_{i=1}^{n}\lambda_i(\tau)\right]d\tau\right\} \qquad (4\text{-}1a)$$

For useful life with all failure rates (λ's) constant

$$R_{ss}(t) = \exp\left[-\left(\sum_{i=1}^{n}\lambda_i\right)t\right] \qquad (4\text{-}1b)$$

Thus the effective failure rate of a system formed from nonredundant components is equal to the sum of the failure rates of the individual components. The components need not be identical.

The mortality of a series system is

$$m_{ss}(t) = \sum_{i=1}^{n}(\lambda_i(t))\exp\left\{-\int_0^t\left[\sum_{i=1}^{n}\lambda_i(\tau)\right]d\tau\right\} \qquad (4\text{-}2)$$

The MTTF of a series system is

$$\bar{T}_{ss} = \int_0^{\infty}\left(\exp\left\{-\int_0^t\left[\sum_{i=1}^{n}\lambda_i(\tau)\right]d\tau\right\}\right)dt \qquad (4\text{-}3)$$

and is dependent on the failure rates of the individual components.

EXAMPLE 4-1   A system consists of four independent components in series, each having a reliability of 0.970. What is the reliability of the system?

$$R_{ss} = \prod_{i=1}^{4} R_i(t) = (0.970)^4 = 0.885$$

If the system complexity is increased so that it contains eight of these components, what is the new reliability?

$$R_{ss} = \prod_{i=1}^{8} R_i(t) = (0.970)^8 = 0.783$$

If the more complex system is required to have the same reliability as the simpler system, i.e., 0.885, what must the reliability of each component be?

$$R_i = (0.885)^{1/8} = 0.985 \qquad ////$$

Reliability is always equal to or less than unity. In addition, it is clear that upon multiplying component reliabilities, the reliability of a complex series system will decrease rapidly with many components. If a system consists of 100 independent components in series (each component having a reliability of 99 percent), the system reliability will be only 36.5 percent. With 400 such components in series, the reliability decreases to 1.8 percent. We conclude that components which are used in large numbers in series in complex systems must have extremely small failure rates.

The MTBF for a series system in useful life can be calculated by converting the individual MTBFs into failure rates, adding these to get a system failure rate and converting this to a system MTBF.

EXAMPLE 4-2 An airborne electronic system has a radar, a computer, and an auxiliary unit with MTBFs of 83, 167, and 500 h, respectively. Find the system MTBF and the reliability for a 5-h operating time.

Unit	MTBF, h	Equivalent failure rate failure/1000 h
Radar	83	12
Computer	167	6
Auxiliary	500	2
		total = 20

System MTBF = 1000/20 = 50 h

For a 5-h mission

$$R_{sys}(5) = e^{-5/50} \simeq 90\%$$

while for radar, $R = 94\%$
computer, $R = 97\%$
auxiliary, $R = 99\%$ ////

4-3 ENERGIZE–DE-ENERGIZE

The time t in Eq. (4-1b) is the system operating time. As written, there is an implied assumption that all components in the system operate continuously for the entire system operating period. In many situations, however, not all components operate in this manner. Rather, they will be *energized* (functioning under load) part of the time and will be *de-energized* (unloaded) the rest of the time. This can be treated in various ways.

One method is to calculate an effective failure rate based on the fraction of time (fraction of operating time) that the component operates under load. If, for example, the component is required to function for one-tenth of the operating period, we can use an effective failure rate of $\frac{1}{10}$ of that given for the component while using the total operating time in Eq. (4-1b).

A second method is to use the given failure rate but to multiply by the actual time (during the total operating time) of actual component operation. This gives the same effective answer.

Under conditions of partial-period operation, a component may be highly reliable in terms of the total operation even though it may have a relatively high failure rate for continuous operation.

It should be noted that both procedures indicated above assume the component has zero failure rate when it is not operating. This assumption of zero failure rate when de-energized (idle, quiescent, etc.) is not always appropriate. In many situations, the idle condition will actually be at low power and some failures should be expected. If a component has a failure rate of λ_e when operating (energized) and a failure rate of λ_d when idle (de-energized) and it operates for t_1 h for every t h of operating time (or $t_2 = t - t_1$ h of idle time), then the effective failure rate is

$$\lambda_{\text{eff}} = \frac{\lambda_e t_1 + \lambda_d (t - t_1)}{t} \qquad (4\text{-}4)$$

Not all components have failure rates expressed in failures per hour. There are many components, such as switches and relays, for which it is more appropriate to express failure rate in terms of failures per cycle. If the failure rate is expressed as failures per cycle of operation and there are c cycles of operation during the operating period, then the effective failure rate for an operating time t is

$$\lambda_{\text{eff}} = \frac{c \lambda_c}{t} \qquad (4\text{-}5)$$

If the component is, for example, a relay which turns on the system at the beginning of the mission, leaves it on for the operating period t, and turns off the system at the end of the mission, there is one cycle. Assuming that λ_e and λ_d of the relay are zero, the relay's contribution to the system failure rate is $1\lambda_c$ and the reliability of the relay for t h is $e^{-1\lambda_c}$.

In many cases, the effect of switching is essentially negligible because the sum of the time-dependent failure rates in a system is usually much larger (and thus much more important) than the per-cycle failure rate of a switching device with few switching operations during the mission. If, however, the switching device performs for a large number of cycles, the system contains a large proportion of switching devices, or various components are sensitive to high-temperature gradients caused by switching (e.g., lamps, tubes, etc.), switching then becomes an important consideration.

A question always arises, i.e., is it better to leave a system on all the time or to switch it on and off when needed? The answer depends on the ratio of the probability of surviving the hours in the unneeded energized condition to the probability of surviving on-off switching cycles. In addition, there is a question of economy, since the cost of energy consumed during unnecessary operation is important. At least equally important is the fact that energized components are subject to wearout or gradual degradation (including exceeding tolerance limits). These processes are nearly nonexistent in the de-energized condition.

EXAMPLE 4-3 Consider a communication system consisting of a transmitter, a receiver, and an encoder. Failure of one causes system failure. During 8 h of communication the units operate for various periods. Determine the reliability for maintaining communications without failure for 8 h.

Unit	Operating period, h	Failure rate, failures/h	λt	R_i
Transmitter	6	0.0267	0.16	0.85
Receiver	8	0.00125	0.01	0.99
Encoder	4	0.015	0.06	0.94
			total = 0.23	

$$R_{\text{sys}} = e^{-\Sigma \lambda t} = e^{-0.23} = 0.79 = 79 \text{ percent} \qquad ////$$

4-4 PARALLEL SYSTEMS

Assume that the n components in Fig. 4-3 are independent and the system will survive for operating time t; if any one of the components survives (or the system fails only if all the components fail), then the unreliability is

$$Q_{SP}(t) = Q_1(t)Q_2(t) \cdots Q_{n-1}(t)Q_n(t)$$

$$\begin{aligned} R_{SP}(t) &= 1 - Q_{SP}(t) \\ &= 1 - [Q_1(t)Q_2(t) \cdots Q_{n-1}(t)Q_n(t)] \\ &= 1 - \{[1 - R_1(t)][1 - R_2(t)] \cdots [1 - R_n(t)]\} \\ &= 1 - \prod_{i=1}^{n} Q_i(t) = 1 - \prod_{i=1}^{n} [1 - R_i(t)] \end{aligned} \qquad (4\text{-}6)$$

(There is an implied assumption that all units are operating simultaneously.)
The mortality of the parallel system is

$$m_{SP}(t) = \sum_{j=1}^{n} m_j(t) \prod_{j \neq i} [1 - R_i(t)] \qquad (4\text{-}7)$$

where the product is taken over all factors i from 1 to n, except when $j = i$.

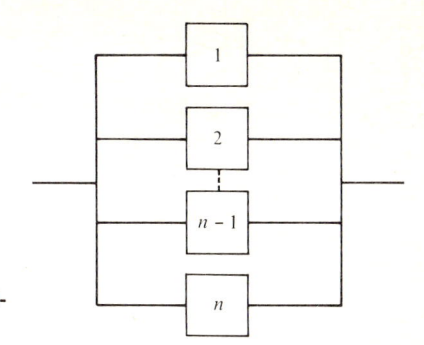

FIGURE 4-3
Schematic system with units in parallel.

The MTBF of the parallel system can be evaluated by applying Eq. (3-3). If the failure rates of all units are identical, then, for this special case,

$$\text{MTBF}_{SP} = \frac{1}{\lambda} + \frac{1}{2\lambda} + \cdots + \frac{1}{n\lambda} \quad (4\text{-}8)$$

A series system (in useful life) has a constant failure rate which is the sum of all the component effective failure rates. Because such a system is exponential, its MTBF is the reciprocal of the (combined) failure rate. *This does not apply to parallel systems.* The instantaneous failure rate of a parallel system is a variable function of the operating time, even though the failure rates and MTBFs of individual components are constant. The MTBF of a parallel (redundant) system is a joint function of the MTBF of each possible path and the number of parallel paths. The MTBF of the system increases slowly as the number of paths increases. The rate of increase is greatest when n (the number of paths) is small. Therefore, the greatest gain in reliability, by adding another path, is realized when a second path is added to a single path. The reliability of a parallel combination is not a simple exponential but a sum of exponentials. *The MTBF cannot be used for determining reliability in a parallel system for an operating time* t *by simple substitution of MTBF in Eq. (3-5)*.

EXAMPLE 4-4 Consider Example 4-3 with a second transmitter added in parallel with the first. Determine the new reliability.

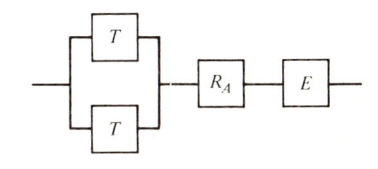

$$R_{\text{trans}} = 1 - (1 - 0.85)^2$$
$$= 0.9775$$
$$R_{\text{sys}} = (0.9775)(0.99)(0.94) = 0.9097 \simeq 91 \text{ percent}$$

Addition of the second transmitter in parallel with the first one increases the reliability from 79 percent to 91 percent. NOTE: The two transmitters would not be considered redundant (and the calculation above would be in error) if failure of one transmitter causes failure of the other, or if the manner of failure of one transmitter prevents the other from assuming the full burden of communication. ////

4-5 COMBINED SERIES-PARALLEL SYSTEMS

Complete systems generally consist of a large number of components or units in series. If any one component or unit fails, the system fails. In many cases, the less reliable components in a system are backed by parallel components to increase system reliability by parallel redundancy. Sometimes a whole group of components is backed by an equal or similar group operating in parallel with the first group. These parallel arrangements of two or more groups can be considered as single units in series within the system. If such a unit fails as a whole, the system fails.

EXAMPLE 4-5 Consider a system in which five different devices must work in series to produce a given result. The reliabilities of the devices are: $A = 95$ percent, $B = 99$ percent, $C = 70$ percent, $D = 75$ percent, and $E = 90$ percent. If the system is composed of five single devices, the reliability is only about 45 percent. This is rather low. Find one arrangement which will give at least 75 percent reliability.

Obviously, the problem is concerned mostly with devices C and D, since the combined reliability of A, B, and E is about 84.5 percent. The lower reliabilities of C and D can be offset by using parallel redundancy. Consider the following possibility.

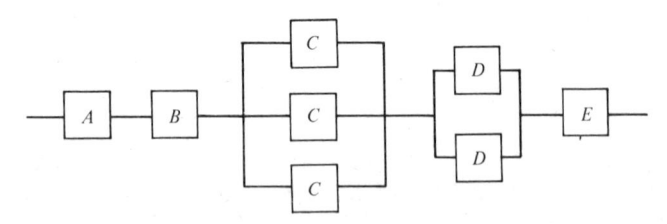

For three C units
$$R = 1 - (1 - 0.70)^3 = 0.973$$

For two D units
$$R = 1 - (1 - 0.75)^2 = 0.9375$$

$$R_{sys} = (0.95)(0.99)(0.973)(0.9375)(0.90) = 0.772$$

It is obvious that other possible arrangements will give at least 75 percent reliability. ////

SERIES AND PARALLEL SYSTEMS   37

EXAMPLE 4-6 Consider a three-element system (each element having a reliability of 0.90). Calculate the reliabilities of the four arrangements given below. Draw appropriate conclusions.

(*a*) *Series*

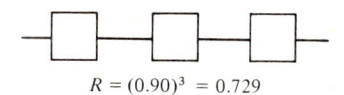

$$R = (0.90)^3 = 0.729$$

(*b*) *Parallel-Series*

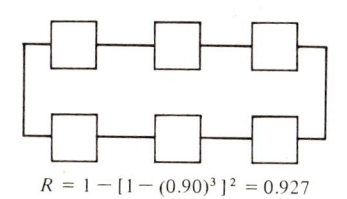

$$R = 1 - [1 - (0.90)^3]^2 = 0.927$$

(*c*) *Mixed-Parallel*

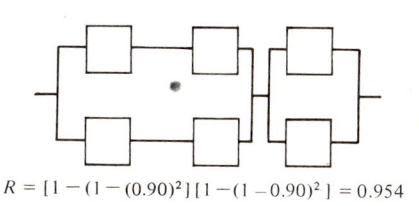

$$R = [1 - (1 - (0.90)^2][1 - (1 - 0.90)^2] = 0.954$$

(*d*) *Series-Parallel*

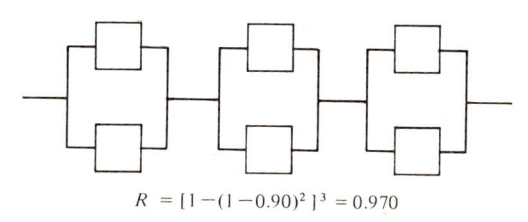

$$R = [1 - (1 - 0.90)^2]^3 = 0.970$$

Conclusions: The reliability of a system with m parallel paths of n elements each (reliability of a single element is p) is

$$R = 1 - (1 - p^n)^m \quad \text{(parallel-series)} \quad (4\text{-}9)$$

The reliability of a system of n series units with m parallel elements (reliability of a single element is p) in each unit is

$$R = [1 - (1 - p)^m]^n \quad \text{(series-parallel)} \quad (4\text{-}10)$$

Series-parallel configurations result in higher reliability than equivalent parallel-series configurations. A basic n-element system replicated m times has m^n possible paths for system success for the series-parallel system; there are only m possible paths in the parallel-series system.

From the foregoing, it follows that the lower the system level at which redundancy is introduced, the more effective redundancy is in increasing reliability.

It is not obvious from the preceding material, but it is true that, as the reliability (p) of the units increases, the difference between parallel-series and series-parallel reliabilities decreases. ////

4-6 BLOCK DIAGRAMS (OR LOGIC DIAGRAMS)

It is very convenient to sketch reliability block diagrams when working on solutions to reliability problems. Typical examples of such diagrams are Figs. 4-2 and 4-3 and the diagrams shown in Examples 4-4, 4-5, and 4-6.

It is necessary to distinguish between the schematic physical diagram and the reliability block diagram. Figure 4-1, for example, is a schematic physical diagram. It is also a reliability block diagram. In many cases, however, the reliability block diagram is not the same as the physical diagram.

The distinction can be made in a somewhat different fashion by considering that the schematic physical diagram shows the *physical* relationships among elements in the system. This is also known as a system diagram. The reliability block diagram (or logic diagram) shows the *functional* relationships among the system elements. The logic diagram consists of groups of blocks, signifying system elements, connected together in a manner which traces the system function.

Consider the schematic physical diagram in Fig. 4-4 of two valves in a pipe line. Would the reliability block diagram appear as a series system or as a parallel system? (After some thought, see the answer at the end of the chapter.)

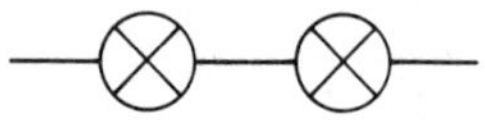

FIGURE 4-4
Schematic diagram of a pipeline with two valves in the line.

4-7 BINOMIAL DISTRIBUTION

In developing Eq. (4-6) (Sec. 4-4), the criterion for system success (Fig. 4-3) required successful functioning of only one of the n components in the system. In many situations, there will be a requirement for successful functioning of more than one of the n components: Calculation of reliability for such a situation can be made by use of the binomial distribution [Eq. (A-25)].

Consider the two switches shown in Fig. 4-1b. Each switch functions properly (p), or it does not (q). Thus $p + q = 1$. The ways in which these two switches can function are:

Result	Probability
Both OK	p^2
One OK, other non-OK	$pq + qp = 2\,pq$
Both non-OK	q^2

This covers all possible ways of functioning. Thus

$$p^2 + 2pq + q^2 = 1$$

But this is $(p + q)^2$.

If we go to a system such as Fig. 4-3 with $n = 3$, the possible ways of functioning are:

Result	Probability
Three OK	p^3
Two OK, one non-OK	$p^2q + pqp + qp^2 = 3p^2q$
One OK, two non-OK	$pq^2 + qpq + q^2p = 3pq^2$
Three non-OK	q^3

Adding

$$p^3 + 3p^2q + 3pq^2 + q^3 = 1$$

But this is $(p + q)^3$.

This can be expanded and generalized to

$$(p + q)^n = p^n + np^{n-1}q + \frac{n(n-1)}{1!2!}p^{n-2}q^2 + \cdots + \frac{n!}{k!(n-k)!}p^{n-k}q^k +$$

$$\cdots + \frac{n(n-1)}{1!2!}p^2q^{n-2} + npq^{n-1} + q^n$$

Rewriting

$$P_n(k) = C_k^n p^{n-k}(1-p)^k \qquad (4\text{-}11)$$

This is the probability of an event not occurring k times in n trials, if the probability of the event in each trial is p. This is also the probability of the event occurring $(n - k)$ times in n trials. The coefficient, C_k^n, can be evaluated by using Eq. (A-11).

EXAMPLE 4-7 What is the probability of two boys in a four-child family if the probability is $\frac{1}{2}$ that a boy will be born at any given birth? (Actually the probability is 0.51.)

$$P_4(2) = \frac{4!}{2!(4-2)!}\left(\frac{1}{2}\right)^2\left(1-\frac{1}{2}\right)^2 = \frac{3}{8} \qquad ////$$

As written, Eq. [4-11] gives the probability of exactly k failures. We can determine the probability of k or fewer failures by using the cumulative distribution function

$$F(k) = \sum_{k=0}^{k} P_n(k) = P(k \text{ or fewer failures}) \qquad (4\text{-}12)$$

To demonstrate applicability to reliability, consider a system such as Fig. 4-3 with three identical units in parallel.

$$(R + Q)^3 = R^3 + 3R^2Q + 3RQ^2 + Q^3$$

		k
R^3	Probability of no failure	0
$3R^2Q$	Probability of one failure	1
$3RQ^2$	Probability of two failures	2
Q^3	Probability of three failures	3

If the requirement is that two out of three must function properly, i.e., none or one can fail,

$$R_{sys} = R^3 + 3R^2Q$$
$$Q_{sys} = 3RQ^2 + Q^3$$
$$R_{sys} = R^3 + 3R^2(1 - R) = 3R^2 - 2R^3$$

If the three units are different, then

$$(R_1 + Q_1)(R_2 + Q_2)(R_3 + Q_3) = 1$$

Expanding

$$R_1R_2R_3 + (R_1R_2Q_3 + R_1Q_2R_3 + Q_1R_2R_3)$$
$$+ (R_1Q_2Q_3 + Q_1R_2Q_3 + Q_1Q_2R_3) + Q_1Q_2Q_3 = 1$$

If the requirement is that two out of three must function properly,

$$R_{sys} = R_1R_2R_3 + (R_1R_2Q_3 + R_1Q_2R_3 + Q_1R_2R_3)$$

Application of the binomial distribution is limited to samples of known size (n is known) with a count of the number of times a certain event is observed.

EXAMPLE 4-8 A certain type of vacuum tube is known to fail prematurely in 3 percent of all cases. If three such tubes are used in three independent sets of equipment (one in each), what is the probability that two of the three will fail prematurely?

$$P_3(1) = C_1^3 p^{(3-1)} q^1$$
$$= \frac{3!}{1!2!}(0.03)^2(0.97)^1$$
$$= 0.2619\%$$

$p = 0.03$
$q = 0.97$
$n = 3$
$k = 1$

("Success" is defined in this problem as premature failure, i.e., the probability of two failing prematurely.)

Probability all three fail prematurely is

$$P_3(0) = C_0^3 p^3 q^0 = 0.0027\%$$

Probability one fails prematurely is

$$P_3(2) = C_2^3 p^1 q^2 = 8.4681\%$$

Probability none fails prematurely is

$$P_3(3) = C_3^3 p^0 q^3 = 91.2673\%$$ ////

4-8 POISSON DISTRIBUTION

The Poisson distribution can be used in a manner somewhat like the binomial distribution, but it is used to deal with events in which the value of n in the expression $(p + q)^n$ is not known, i.e., sample size is indefinite.

The Poisson distribution obtains an expression equivalent to $(p + q)^n = 1$ from

$$e^{-x} e^x = 1$$

Expanding e^x

$$e^{-x}\left(1 + x + \frac{x^2}{2!} + \cdots + \frac{x^n}{n!} + \cdots\right) = 1$$

This distribution is defined over a sample space which is countably infinite. The interpretation is:

If x is taken to be the expected or average number of occurrences of an event, then

e^{-x} = the probability that the event will not occur
xe^{-x} = the probability that the event will occur exactly once
$(x^2/2!)e^{-x}$ = the probability that the event will occur exactly twice
$(x^n e^{-x})/n!$ = the probability that the event will occur exactly n times

In reliability, the event is failure and the average number of failures in time t is given by λt when the failure rate is constant. Consequently,

$e^{-\lambda t}$ is $R(t)$ for a system having a constant failure rate λ. This is the probability that no failure will occur in time t.

$(\lambda t)e^{-\lambda t}$ is the probability of exactly one failure occurring in time t.

$[(\lambda t)^2/2!]e^{-\lambda t}$ is the probability of exactly two failures occurring in time t.

In general, the probability of exactly k failures occurring in time t is given by

$$\frac{e^{-\lambda t}(\lambda t)^k}{k!} \tag{4-13}$$

The cumulative Poisson distribution (the probability of k or fewer failures) is given by

$$\sum_{k=0}^{k} \frac{e^{-\lambda t}(\lambda t)^k}{k!} \tag{4-14}$$

This distribution can be used to determine the probability of some specified number of failures during a given mission or to calculate the number of spares required when units operate in useful life.

4-9 MULTINOMIAL DISTRIBUTION

The binomial distribution, Sec. 4-7, provides a very useful tool in making reliability calculations. It has a limitation, however, since it applies only to "go–no go" situations. There are situations in which there may be more than two ways for a unit to operate, i.e., it may perform properly and it may malfunction in two or more ways. Such cases can be handled by using the multinomial distribution, which is essentially a generalization of the binomial.

Let n be the number of independent trials, in each of which only one of the events $A_1, A_2, \ldots, A_m$ can occur. The probabilities of these events in each of the trials are $p_1, p_2, \ldots, p_m$, respectively. The probability that A_1 will occur k_1 times, A_2 will occur k_2 times, $\ldots$, and A_m will occur k_m times in n trials is

$$P_n(k_1, k_2, \ldots, k_m) = \frac{n!}{k_1! k_2! \cdots k_m!} p_1^{k_1} p_2^{k_2} \cdots p_m^{k_m} \quad (4\text{-}15)$$

with the restrictions that $\sum_{i=1}^{m} k_i = n$

and $p_1 + p_2 + \cdots + p_m = 1$

EXAMPLE 4-9 Consider a diode which can function properly but can malfunction by short-circuiting or by open-circuiting. Let the probabilities of these be

p_n = probability of proper operation
p_s = probability of short-circuiting
p_o = probability of open-circuiting

(Normally we would expect p_n very close to 1 with p_s and p_o very small.) It is certain that the diode will do one of these three; therefore, $p_n + p_s + p_o = 1$.

If we consider four identical diodes for improving the reliability, these can be arranged in two possible configurations as shown. What is the improvement in using a diode quad? (Criterion for system success is flow of a controlled quantity of electrical current.)

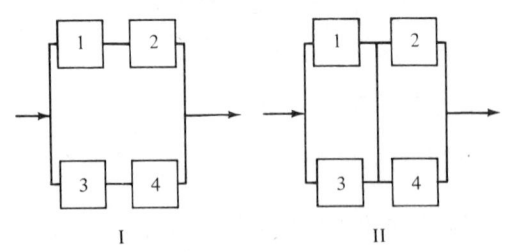

Applying the multinomial theorem [Eq. (4-13)] and expanding

$$(p_n + p_s + p_o)^4 = p_n^4 + 4p_n^3 p_s + 4p_n^3 p_o + 6p_n^2 p_s^2 + 12p_n^2 p_s p_o + 6p_n^2 p_o^2$$
$$+ 4p_n p_s^3 + 12p_n p_s^2 p_o + 12p_n p_s p_o^2 + 4p_n p_o^3 + p_s^4 + 4p_s^3 p_o$$
$$+ 6p_s^2 p_o^2 + 4p_s^3 p_o + p_o^4 = 1$$

SERIES AND PARALLEL SYSTEMS   43

This expansion automatically gives all possible permutations of the four diodes among the three possible modes of function, independently of configuration. The next question is that of satisfactory function of a configuration, each of which must be considered separately. Considering configuration I, let us examine the expansion, term by term.

p_n^4 — all normal, system success

$4p_n^3 p_s$ — three normal, one short-circuited, system success—any single diode can be short-circuited with quad operating properly; all four terms apply

$4p_n^3 p_o$ — three normal, one open-circuited; system success—if only one is open-circuited, the path containing it is out, but the other path will function properly; all four terms apply

$6p_n^2 p_s^2$ — two normal, two short-circuited; if diodes 1 and 2 or 3 and 4 are short-circuited, there is a direct short-circuit and the quad cannot function. If there is one short-circuit in each branch (there are four ways of doing this: 1-3, 1-4, 2-3, and 2-4), the quad can function; therefore, four terms are successes and two terms are failures.

Each remaining term of the expansion can be examined and the following table developed for configuration I.

Success	Failure
p_n^4	
$4 p_n^3 p_s$	
$4 p_n^3 p_o$	
$4 p_n^2 p_s^2$	$2 p_n^2 p_s^2$
$12 p_n^2 p_s p_o$	
$2 p_n^2 p_o^2$	$4 p_n^2 p_o^2$
	$4 p_n p_s^3$
$8 p_n p_s^2 p_o$	$4 p_n p_s^2 p_o$
$4 p_n p_s p_o^2$	$8 p_n p_s p_o^2$
	$4 p_n p_o^3$
	p_s^4
	$4 p_s^3 p_o$
	$6 p_s^2 p_o^2$
	$4 p_s p_o^3$
	p_o^4

R = system reliability = probability of quad success
Q = system unreliability = probability of quad failure

$$R = p_n^4 + 4p_n^3 p_s + 4p_n^3 p_o + 12 p_n^2 p_s p_o + 4p_n^2 p_s^2 \\ + 2p_n^2 p_o^2 + 8 p_n p_s^2 p_o + 4 p_n p_s p_o^2$$

$$Q = 2p_n^2 p_s^2 + 4 p_n^2 p_o^2 + 4 p_n p_s^3 + 4 p_n p_s^2 p_o + 8 p_n p_s p_o^2 \\ + 4 p_n p_o^3 + p_s^4 + 4 p_s^3 p_o + 6 p_s^2 p_o^2 + 4 p_s p_o^3 + p_o^4$$

Eliminating p_n by substitution, i.e., $p_n = (1 - p_s - p_o)$ and collecting terms,

$$Q = 2p_s^2 + 4p_o^2 - 4p_o^3 + p_o^4 - p_s^4$$

If p_s and p_o are small, the last three terms can be neglected, then

$$Q \simeq 2p_s^2 + 4p_o^2$$

The improvement factor is

$$G \equiv \frac{Q \text{ for single}}{Q \text{ for quad}} = \frac{p_s + p_o}{2p_s^2 + 4p_o^2}$$

If $p_s = p_o = 0.01$, $G_I = 33.33$ ////

4-10 SOLUTION TO 4-6

A question was asked whether the reliability block diagram for Fig. 4-4 should be shown as series or parallel. The proper answer depends on the definition or criterion of adequate performance of the system. If the two valves are normally shut but are expected to open on command to provide flow, this is a series system in terms of reliability. If, however, the two valves are normally open but are expected to shut on command to stop flow, this is a parallel system in terms of reliability.

PROBLEMS

4-1 A system consists of 100 units in series, each unit having a reliability of 0.99. What is the reliability of the system? What fraction of such systems will perform satisfactorily?

4-2 An old-fashioned string of Christmas tree lights has 10 bulbs connected in series. What must the reliability of each bulb be if there is to be a 90 percent chance of the string lighting after one year of storage?

4-3 An unmanned missile designed for space exploration has 1000 components in series. If the mission is required to have a reliability of 90 percent and each component has the same reliability, what must be the reliability of each component?

4-4 A given component has a constant failure rate of 0.0050 failure/h. Determine:
   (a) Reliability of one unit for an operating time of 300 h.
   (b) Reliability of two units in series for an operating time of 150 h.
   (c) Reliability of three units in series for an operating time of 100 h.
   (d) Reliability of four units in series for an operating time of 75 h.

4-5 A system has n nonredundant components; their failure rates λ_i are constant, but (possibly) all different. Find the reliability, mortality, and MTTF of the system. Specialize for the case when all λ_i are equal.

4-6 Find the MTBF of the system if five components in series have constant failure rates of 1.4, 1.7, 1.9, 1.2, and 1.6 failures per 1000 h, respectively.

4-7 An equipment has 585 components, as listed below. It is specified that the equipment must have a reliability of at least 95 percent for an 18-h mission time. Does the equipment meet this specification? If so, how well?

Component type	No. in use	Base failure rate, %/1000 h
Transistors	100	0.1
Diodes, crystal	200	0.03
Connectors	5	0.2
Resistors	150	0.1
Capacitors	100	0.1
Switch	1	0.5
Relay	1	0.5
Sockets	6	0.3
Potentiometers	20	0.5
Transformers	2	0.2
Total	585	

4-8 A device is cycled on and off during routine operation. It has a failure rate of λ_e failure/h while functioning, a failure rate of λ_d failure/h while idling, a switch sensing failure rate of λ_{se} failures/h, a switch failing-open failure rate of λ_{so} failures/h, and a switching failure rate of λ_c failures/cycle. Derive an average failure rate for the device.

4-9 Consider three identical units in parallel. What is the system reliability if each unit has a reliability of 95 percent?

4-10 A system consists of five identical components connected in parallel. What must be the reliability of each component, if the reliability of the system must be 98 percent?

4-11 If a system must have reliability of 99 percent, how many components are required in parallel if each component has a reliability of 60 percent?

4-12 A system consists of five components connected as shown. Find the overall reliability of the system.

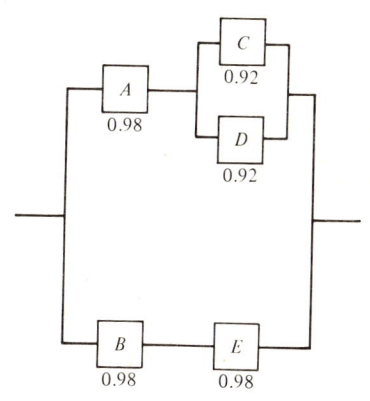

4-13 You have a choice of the two designs shown. Elements A and B have MTBFs of 15 and 20 h, respectively. Which design would you choose if mission time is (a) 5 h, (b) 15 h, (c) 30 h? What is the reaction to using either design under the limited conditions indicated?

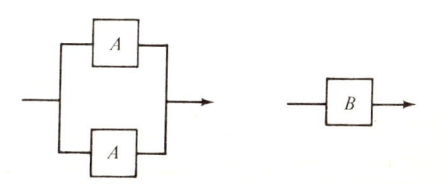

46 INTRODUCTION TO RELIABILITY IN DESIGN

4-14 Determine the reliability and the unreliability of the circuit shown below.

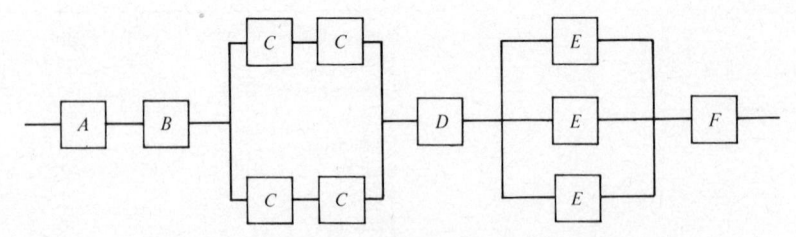

4-15 A system designer has the choice of the three configurations shown below with individual element reliabilities as indicated. Assume independence of elements. Which configuration do you recommend as the most reliable? Show calculations to support your recommendation.

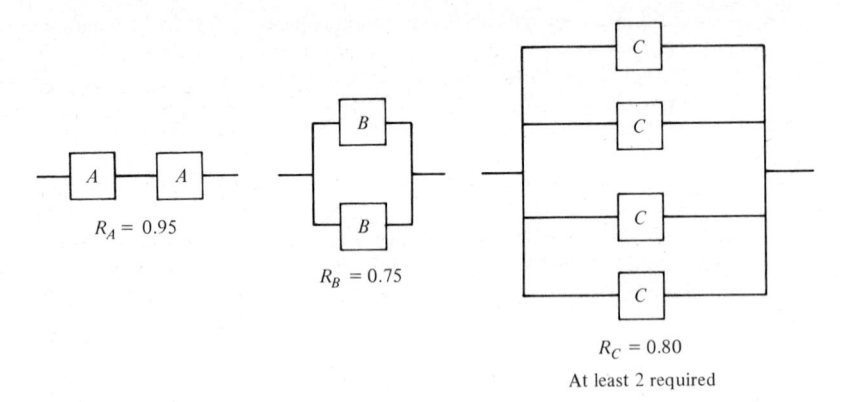

$R_A = 0.95$

$R_B = 0.75$

$R_C = 0.80$
At least 2 required

4-16 Three systems, as shown, are in competition. Which would you select if the major criterion were reliability? If each of the three systems shown will perform the desired function equally effectively, which offers the highest reliability per dollar if the costs of individual elements are A: \$1000, B: \$500; and C: \$200?

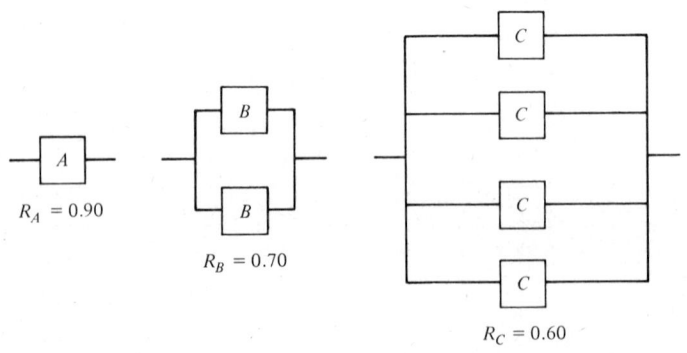

$R_A = 0.90$

$R_B = 0.70$

$R_C = 0.60$
At least two must function

4-17 An airborne system consists of three black boxes. These may be arranged in any one of the four following configurations.

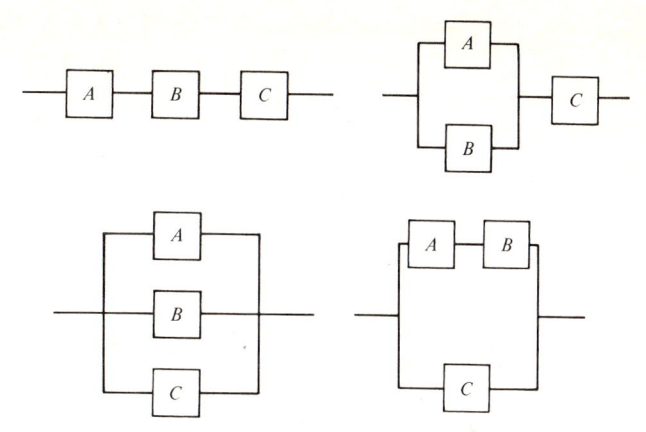

The reliabilities are: $R_A(t) = e^{-\alpha t}$; $R_B(t) = e^{-\beta t}$; and $R_C(t) = e^{-\gamma t}$.
(a) Determine the reliability of each configuration.
(b) If $\alpha = \beta = \gamma$, compare the system reliabilities.

4-18 A system has three units with failure rates of 4.0, 15, and 3.8 failures/10^6 h, respectively. Find the MTBF and the reliability for both 100 h and 10 h for (a) the three units in series, and (b) the three units in parallel.

4-19 A chemical processing system is equipped with a pressure controller, three temperature controllers, and two level alarms, the latter being operated in parallel for safety. The plant will fail if the pressure, level, or any one temperature is not effectively controlled. The mean failure rate of each instrument is given. For a period of continuous operation of one week, calculate the probability of system failure due to ineffective instrumentation. (It might be helpful to show the probability of failure of each type of instrument.)

Instrument	Failure rate, %/1000 h
Pressure controller	60
Temperature controller	12
Level alarm	83

4-20 A nonredundant subsystem has 200 elements in it, each element having an MTBF of 0.5×10^6 year. In order to obtain a system MTBF of 1 year, it is necessary to have 19 such subsystems in parallel. This obviously is relatively complex, bulky, and expensive. The system producer finds a supplier making elements which have essentially the same performance characteristics, but the individual elements have an MTBF twice that of the original elements. The new elements cost four times as much as the original elements. Considering the cost of the elements (neglecting assembly costs, etc.), which elements would you use? Why?

4-21 Consider the case of three elements, as shown in the diagram below. At least two of the three elements must function properly for the system to function properly. All three elements are different, having reliabilities of p_a, p_b, and p_c, respectively. Find the reliability of the system.

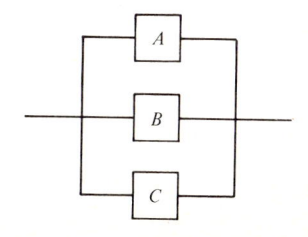

4-22 A delivery company (such as United Parcel Service) has a fleet of trucks equipped with tires having a failure rate of 4×10^{-6} failure/mile. Two types of trucks are used: one type with four tires; the other type with six tires, i.e., two on each end of the rear axle. The same tires are used on both types of truck. Sketch a reliability block diagram for each type of truck, and calculate the probability of each type being unable to complete deliveries due to tire trouble in 10,000 miles of driving.

4-23 An operator of a fleet of trucks has terminals in Detroit and Chicago. The probability that any one truck will have tire, engine, or other trouble resulting in lost time on one run is 0.05. A major Detroit customer has a large order of perishables which he wishes delivered in Chicago without delay. Delivery of this order will require use of 12 trucks.
(a) What is the reliability for immediate delivery of the complete order?
(b) What is the probability that no more than one truck will be delayed?

4-24 A delivery company (such as United Parcel Service, which functions nationwide) has a fleet of trucks equipped with tires having a failure rate of 4×10^{-6} failure/mile. Three types of trucks are used: a van type with one tire per wheel; a heavier truck with dual tires on the rear wheels; and a semitrailer with dual tires on the rear wheels of the tractor and dual tires on the single axle of the trailer. The same tires are used on all three types of truck. Sketch a reliability block diagram for each type of truck and calculate the reliability for each type being able to complete deliveries without tire trouble in 25,000 miles of driving. What conclusions do you draw from comparing the three?

4-25 The probability that a car driving the entire length of a certain turnpike will have a blowout is 0.05. Find the probability that, among 18 cars traveling the length of this turnpike, (a) exactly one will have a blowout, (b) at most three will have a blowout, and (c) two or more will have a blowout.

4-26 Determine the reliability for the system shown when (a) two out of five units (7 through 11); or (b) three out of five units (7 through 11) are needed for successful operation. What is the reliability of the system if unit 3 or 4 and unit 5 or 6 must function plus four out of five of the remaining units (7 through 11)?

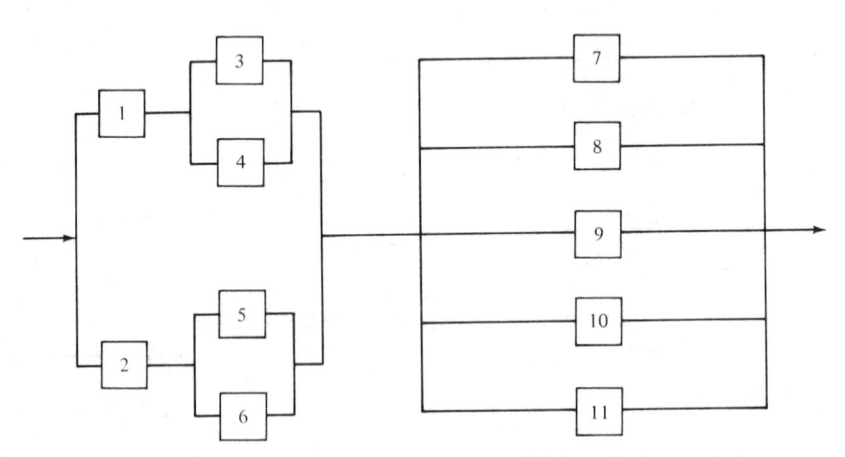

4-27 A device has a random failure rate of 20 failures/10^5 h. For an operating period of 300 h, what is the probability of (a) no failure, (b) one failure, (c) two failures, (d) two failures or less, or (e) more than two failures?

4-28 A system has a constant failure rate of 200×10^{-6} failures/h. For a mission time of 100 h, what is the probability of (a) no failures, (b) one failure, (c) two failures, or (d) three or more failures?

4-29 A given system must have a reliability of at least 95 percent for an operating period of 300 h. The system has one unit which is crucial and must have a reliability of at least 98.5 percent to give the required system reliability. The unit is not that reliable. Thus, it is necessary to have spares available which can be used for immediate replacement upon failure. If the unit MTBF is 1200 h, how many spares are needed per operating period (on the average)?

4-30 The problem of diode quads was discussed in Example 4-9. It was noted that there are at least two possible configurations for such a quad. Configuration I was analyzed and an improvement factor was determined. Analyze configuration II to determine reliability and improvement factor. Under what conditions, if any, is quad I preferable? Under what conditions, if any, is quad II preferable?

4-31 Consider the circuit shown. The reliability of each individual relay is p (for closing and functioning properly). The failure mode is failing open-circuited. Assume that all relays function independently. What is the reliability of the system shown (in terms of p) if the definition of system success is that current flows from terminal L to terminal R? If $p = 95$ percent, what is the numerical value of the reliability?

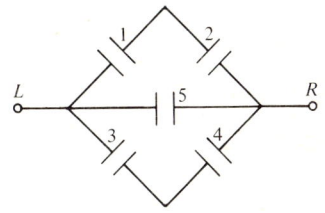

4-32 All details are the same as in Prob. 4-31, except that the configuration differs as shown. Compare with Prob. 4-31. Which would you choose?

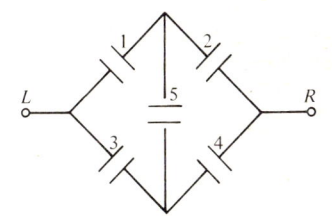

5
STANDBY SYSTEMS

5-1 INTRODUCTION

The basic model of parallel redundancy discussed in Sec. 4-4 has two assumptions which are not always justified. One is the assumption that the system fails only after all the components have failed. This was treated in one form by using the binomial distribution. The second assumption is that the reliabilities of the components are independent.

Consider the situation of a number of generators, in parallel, supplying a common load, as shown in Fig. 5-1. Assume the load is 100 kW and each generator is capable of supplying a maximum of 25 kW. It is obvious that there must be a minimum of four generators in parallel to supply the load. In this case, system failure occurs not when all the generators have failed, but when more than $(n-4)$ have failed. The problem of determining reliability in a system with n components in parallel redundancy, if at least k components are needed for successful operation (or the system fails if more than $n-k$ components fail), can be solved by using the binomial, Poisson, or multinomial distributions. Obviously, this includes the special case of $k=1$, discussed in Sec. 4-4.

If, in the above example, the initial number of generators is greater than

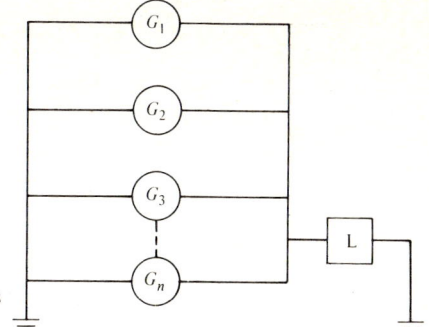

FIGURE 5-1
Schematic diagram of n generators supplying a common load.

four, each generator will be working at less than its maximum capacity. As successive generators fail, the remaining generators will be working under increased loading, even if there are enough of them to supply the full 100 kW. This occurs because the quantity W_a/W_r, where W_a is the actual power supplied and W_r is the rated power, is a stress variable. Any device working at full load has a higher failure rate than one working at a fraction of its capacity. The reliabilities of individual devices in a parallel system, therefore, are generally not independent. If one component fails, the others work under increased stress and their reliabilities decrease. This is especially true of electrical equipment.

When requirements for high reliability make redundancy necessary, a good arrangement has one unit operating until it fails. At that time, a second unit which has been idly standing by is switched into the system.

5-2 STANDBY MODEL

Consider the system, shown in Fig. 5-2, in which component C_1 is operating. There are a number of spare components (or subsystems) which can be automatically switched to take over the system requirements when the operating

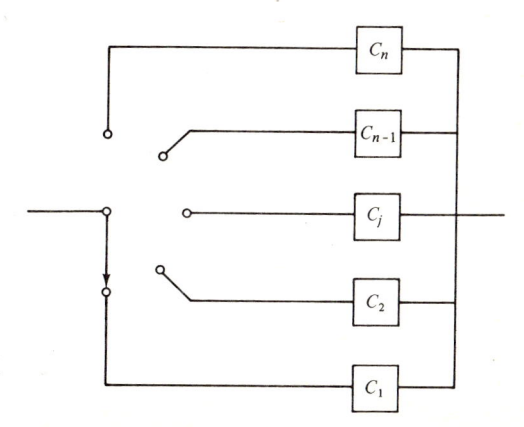

FIGURE 5-2
Schematic diagram of standby system with one unit functioning to supply load.

component fails. Thus, when C_1 fails, C_2 is switched in to take its place. When C_2 fails, C_3 is switched in to take its place, and so on until C_n is supplying the system requirements. The entire system does not fail until C_n fails. The system had one component originally operating with $(n-1)$ standbys. The standbys do not operate until their individual turns come to replace the previously operating component. These standbys are also known as *cold* spares.

A standby system thus has units which stand by idly until they are called on to operate by a sensing and switching subsystem, in contrast to a parallel system where all units operate simultaneously.

5-3 TWO-UNIT STANDBY SYSTEM

A two-unit standby system (Fig. 5-3) functions successfully when the functioning unit does not fail (Fig. 5-3a), *or* if the functioning unit fails during operating time *t*, the sensing and switching unit functions properly, and the standby unit (not having failed while idle) functions properly for the remainder of the mission (Fig. 5-3b). In probabilistic terms: the reliability of the system is the probability that unit 1 succeeds for the whole period (t) *or* that unit 1 fails at some time t_1 prior to t *and* the sensing and switching unit does not fail by t_1 and standby unit 2 does not fail by t_1 *and* successfully functions for the remainder of the mission.

This can be written (assuming 100 percent reliability of the sensing and switching unit and of unit 2 while idling) as

$$R_{SB}(t) = R_1(t) + Q_1(t_1)R_2(t - t_1) \tag{5-1}$$

where $t_1 < t$. This is shown graphically in Fig. 5-4.

The time t_1 can be any value from zero (immediate failure of unit 1) to t (no failure of unit 1). Expansion of Eq. (5-1) gives (for useful life)

$$R_1(t) = e^{-\lambda_{e_1} t}$$

$$Q_1(t_1) = \int_{t_1=0}^{t_1=t} m_1(t_1)\, dt_1$$

From Eq. (3-6).

$$m_1(t_1) = \lambda_{e_1} e^{-\lambda_{e_1} t_1}$$
$$R_2(t - t_1) = e^{-\lambda_{e_2}(t - t_1)}$$

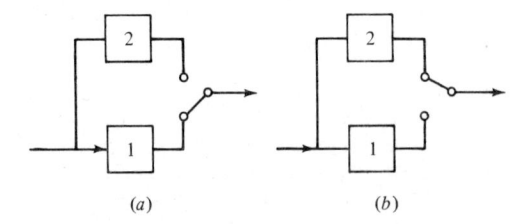

FIGURE 5-3
Reliability block diagram of two-unit standby system.

(a)   (b)

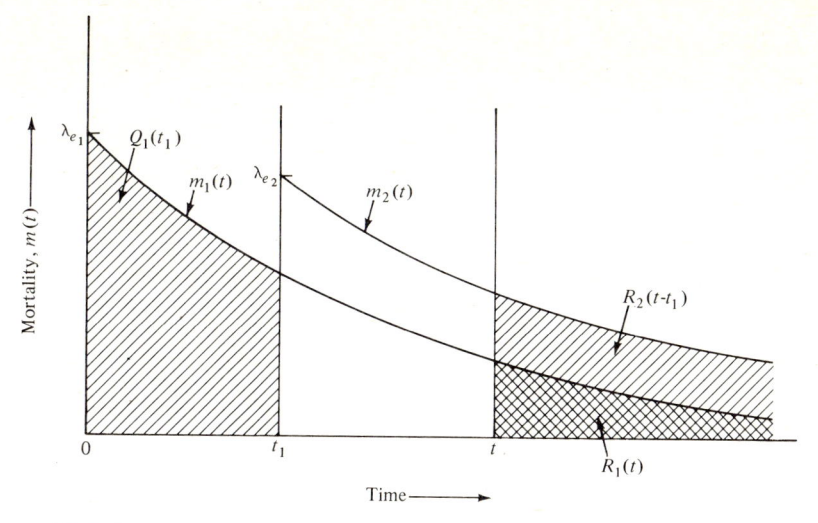

FIGURE 5-4
Graphical representation of terms in Eq. (5-1) in useful life.

$$R_{SB}(t) = e^{-\lambda_{e_1}t} + \left(\int_{t_1=0}^{t_1=t} \lambda_{e_1}e^{-\lambda_{e_1}t_1}\,dt_1\right)(e^{-\lambda_{e_2}(t-t_1)})$$

$$R_{SB}(t) = e^{-\lambda_{e_1}t} + \lambda_{e_1}e^{-\lambda_{e_2}t}\left(\int_{t_1=0}^{t_1=t} e^{-(\lambda_{e_1}-\lambda_{e_2})t_1}\,dt_1\right)$$

$$R_{SB}(t) = e^{-\lambda_{e_1}t} + \frac{\lambda_{e_1}e^{-\lambda_{e_2}t}}{\lambda_{e_1} - \lambda_{e_2}}(1 - e^{-(\lambda_{e_1}-\lambda_{e_2})t}) \tag{5-2}$$

If $\lambda_{e_1} = \lambda_{e_2} = \lambda$, then

$$R_{SB}(t) = e^{-\lambda t} + (\lambda t)e^{-\lambda t} \tag{5-2a}$$

The case of the two-unit standby system with the same failure rate for both units can be viewed as a situation in which the probability of system success is the probability of one failure or less. Using the Poisson distribution, Eq. (4-14), the reliability is

$$R(t) = \sum_{k=0}^{1} \frac{e^{-\lambda t}(\lambda t)^k}{k!}$$

$$R(t) = e^{-\lambda t} + (\lambda t)e^{-\lambda t} \tag{5-2a}$$

We can determine the MTTF for a two-unit standby system by applying Eq. (3-3). For the case of a system in useful life with both units having the same failure rate,

$$\bar{T} = \int_0^\infty R_{SB}(t)\,dt = \int_0^\infty e^{-\lambda t}\,dt + \int_0^\infty (\lambda t)e^{-\lambda t}\,dt$$

$$= \frac{1}{\lambda} + \frac{\lambda}{\lambda^2} = \frac{2}{\lambda} \tag{5-3}$$

Thus the MTTF of the system as a whole is double that for a single unit [Eq. (3-3a)]. It is also greater than the MTTF of a system with two units in parallel [Eq. (4-8)]. Like that of the parallel system, however, the failure rate of a standby system is not constant. Therefore, reliability is not exponential and must be calculated for each case. The MTTF must also be calculated for each case.

EXAMPLE 5-1 Calculate the reliability for a 10-h operating period of a parallel system with two units, each having a failure rate of 0.01 failure/h. Do likewise for a two-unit standby system using the same units; assume 100 percent reliability of sensing, switching, and idling. Compare the two on the basis of reliability and MTBF.

For a single unit [Eq. (4-1)]

$$R(10) = e^{-\lambda t} = e^{-0.10} = 0.90484$$

For a parallel system [Eq. (4-6)]

$$R_{SP}(10) = 1 - (1 - 0.90484)^2 = 0.990945$$

From Eq. (4-8)

$$\text{MTBF} = \frac{1}{\lambda} + \frac{1}{2\lambda} = 150 \text{ h}$$

For a standby system [Eq. (5-2a)]

$$R_{SB}(10) = e^{-\lambda t}(1 + \lambda t)$$
$$= 0.90484(1 + 0.1) = 0.995324$$

From Eq. (5-3)

$$\text{MTBF} = \frac{2}{\lambda} = 200 \text{ h}$$

In this case, it is obvious that there is relatively little more reliability for the standby than for the parallel system despite a clearly greater MTBF. This is true, in general.
////

Two simplifying assumptions were made before writing Eq. (5-1). Both may be valid, but there are occasions when they are not justified. In these latter situations, a more realistic equation for reliability is

$$R_{SB}(t) = R_1(t) + \int_{t_1=0}^{t_1=t} [m_1(t_1) \, dt_1] R_{SSw}(t_1) R_2(t_1) R_2(t - t_1) \qquad (5\text{-}4)$$

If all the components of the system are in useful life, unit 1 has an energized failure rate of λ_{e_1}, the sensing element has a failure rate of λ_{se}, the switching element has a failure rate of λ_{sw} and operates for one cycle, standby unit 2 has a de-energized failure rate of λ_{d_2}, and an energized failure rate of λ_{e_2}, then

$$R_{SB}(t) = e^{-\lambda_{e_1} t} + \int_{t_1=0}^{t_1=t} \lambda_{e_1} e^{-\lambda_{e_1} t_1} e^{-\lambda_{se} t_1} e^{-\lambda_{sw} \times 1} e^{-\lambda_{d_2} t_1} e^{-\lambda_{e_2}(t-t_1)} dt_1$$

$$R_{SB}(t) = e^{-\lambda_{e_1} t} + \lambda_{e_1} e^{-\lambda_{sw}} e^{-\lambda_{e_2} t} \int_{t_1=0}^{t_1=t} e^{-(\lambda_{e_1}+\lambda_{se}+\lambda_{d_2}-\lambda_{e_2}) t_1} dt_1$$

$$= e^{-\lambda_{e_1} t} + \frac{\lambda_{e_1} e^{-\lambda_{sw}} e^{-\lambda_{e_2} t}}{\lambda_{e_1} + \lambda_{se} + \lambda_{d_2} - \lambda_{e_2}} [1 - e^{-(\lambda_{e_1}+\lambda_{se}+\lambda_{d_2}-\lambda_{e_2}) t}] \quad (5\text{-}5)$$

For the special case where $\lambda_{sw} = \lambda_{se} = \lambda_{d_2} = 0$, this reduces to Eq. (5-2) or its alternative form

$$R_{SB}(t) = e^{-\lambda_{e_1} t} + \frac{\lambda_{e_1}}{\lambda_{e_1} - \lambda_{e_2}} [e^{-\lambda_{e_2} t} - e^{-\lambda_{e_1} t}] \quad (5\text{-}2)$$

5-4 MULTIUNIT STANDBY SYSTEM

With the assumption of 100 percent reliability of the sensing-switching unit and the standby units while idling, Eq. (5-2a) can be expanded to the situation having n identical units with one functioning and $(n - 1)$ units in standby to give

$$R_{SB}(t) = e^{-\lambda t} \left[1 + \lambda t + \frac{(\lambda t)^2}{2!} + \cdots + \frac{(\lambda t)^{n-1}}{(n-1)!} \right] \quad (5\text{-}6)$$

Likewise, Eq. (5-3) can be expanded to

$$\bar{T}_{SB} = \text{MTTF}_{SB} = n/\lambda = n\bar{T} \quad (5\text{-}7)$$

If we let X_j be the life of the jth unit (its age at failure, measured from the moment it started operation, not from the time it started as a standby unit), then the time of failure T_j of the jth unit, counted from the moment the system went into operation, is

$$\bar{T}_j = X_1 + X_2 + \cdots + X_j \quad (5\text{-}8)$$

Since the system fails only when all n components have failed, the time to failure of the system is

$$\bar{T}_{SB} = \bar{T}_n = \sum_{j=1}^{n} X_j \quad (5\text{-}9)$$

Mortality, by definition (Sec. 3-1), is the probability density of the time to failure T. It is, therefore, the a priori probability that the system will fail in the time interval dt. For a system having n units in standby (Fig. 5-2), the above applies if $(n - 1)$ units fail in the interval between 0 and t and the nth unit fails in the interval dt. Since failures of cold spares are independent events, we have

$$m(t) dt = P(t < T < t + dt) = P_t(n - 1) P_{dt}(1) \quad (5\text{-}10)$$

where $P_t(n - 1)$ is the probability that $(n - 1)$ units will fail between 0 and t, and $P_{dt}(1)$ is the probability that one unit (the last one) will fail between t and $t + dt$.

If the failure rate of all units is constant and equal to λ, then the two probabilities are given by the Poisson distribution:

$$P_t(n-1) = \frac{(\lambda t)^{n-1}}{(n-1)!} e^{-\lambda t}$$

$$P_{dt}(1) = (\lambda\, dt) e^{-\lambda dt}$$

Upon substituting,

$$m(t)\, dt = \frac{(\lambda t)^{n-1}(\lambda\, dt) e^{-\lambda t} e^{-\lambda dt}}{(n-1)!}$$

Since $dt \ll t$, $e^{-\lambda dt} \simeq 1$, relatively.
The mortality then becomes

$$m(t) = \frac{\lambda^n t^{n-1} e^{-\lambda t}}{(n-1)!} \qquad (5\text{-}11)$$

which is the gamma distribution. (See Appendix A.)
The system reliability [Eq. (2-13)] is

$$R(t) = \int_t^\infty m(\tau)\, d\tau = \frac{\lambda^n}{(n-1)!} \int_t^\infty \tau^{n-1} e^{-\lambda\tau}\, d\tau$$

Letting $\lambda\tau = u$ and substituting

$$R(t) = \frac{1}{(n-1)!} \int_{\lambda t}^\infty u^{n-1} e^{-u}\, du$$

We know that

$$\int_0^\infty u^{n-1} e^{-u}\, du = \Gamma(n)$$

We also know that

$$\int_{\lambda t}^\infty u^{n-1} e^{-u}\, du = \Gamma(n, \lambda t) = \Gamma(n) - \gamma(n, \lambda t)$$

where

$$\gamma(n, \lambda t) = \int_0^{\lambda t} u^{n-1} e^{-u}\, du$$

is the incomplete gamma function. Thus

$$R(t) = \frac{\Gamma(n, \lambda t)}{\Gamma(n)} = 1 - \frac{\gamma(n, \lambda t)}{\Gamma(n)} \qquad (5\text{-}12)$$

The system-failure rate [Eq. (3-2)] is

$$\lambda_{\text{sys}}(t) = \frac{m(t)}{R(t)} = \frac{t^{n-1} e^{-\lambda t}}{\int_t^\infty \tau^{n-1} e^{-\lambda\tau}\, d\tau}$$

$$= \frac{t^{n-1} e^{-\lambda t}}{\Gamma(n, \lambda t)} \qquad (5\text{-}13)$$

This very obviously is a function of time. This system, composed of nonaging components with constant failure rates, is itself aging, since its failure rate increases with time. No contradiction is involved. This can be resolved by visualizing a large number (N) of such systems. There is a total of nN components failing at a constant rate λ. Initially, this leaves most systems intact, since they will have unused spares and the rate of failure of the systems will be very low. As time increases, the constant failure rate of the components will use up spares. When the time is about equal to nT, i.e., when many systems are operating on their last spare, the failure rate of the systems will be high. Instantaneous system-failure rate is a function of time therefore, even though failure rates of individual components are not.

The MTBF of the system is

$$\bar{T}_{SB} = n\bar{T} \qquad (5\text{-}7)$$

if all components have the same failure rate λ.

The MTBF for one component is

$$\bar{T} = \frac{1}{\lambda} \qquad (3\text{-}3a)$$

In the preceding discussion, the Poisson distribution and the gamma distribution were used to describe the same process of a random succession of failures. The Poisson distribution gives the probability that n failures will occur in a given time interval t, whereas the gamma distribution gives the probability that the time necessary for a given number n of failures will be near t. The Poisson distribution is the distribution of the discrete random variable n, while the gamma distribution is the probability density of the continuous random variable t. These two distributions, therefore, describe the same process in terms of different random variables.

5-5 SOME SUMMARY COMMENTS

The reliability of a system with standby redundancy using spares can be made arbitrarily close to 100 percent by providing a sufficiently large number of cold spares. This implies a system which is not subject to maintenance during operation, e.g., a satellite. The number of spares actually used is subject to such constraints as maximum permissible weight, size, and/or cost.

All the discussion in this chapter has essentially dealt with the case of constant failure rates. This is the most important case for a system with standby redundancy. If the reliability of a given system is sufficiently important to justify the relatively expensive procedure of using a sensing-switching subsystem to incorporate cold spares, it would (generally) be inconsistent to use components with increasing failure rates. Decreasing failure rates are unreasonable, since reliability could be increased by aging the components before putting them into service.

One might infer from the discussion in this chapter that standby reliability is limited to the case where all units have the same failure rate. This is not true. It often occurs that the standby units differ from the initial operating unit. A pneumatic unit, for example, may have a hydraulic unit for backup. For such situations, there is a general standard approach. The mortality function of the combination of units is first derived. The system reliability can then be found by applying Eq. (2-13) and integrating. The MTTF can then be found by applying Eq. (3-3) and integrating. This procedure is always permissible. It is always correct. It is independent of the behavior of the components, i.e., whether they are in useful life with constant failure rates or fail in some other manner.

EXAMPLE 5-2 Consider the system shown. If unit 1 fails, unit 2 begins to function. If unit 2 fails, unit 3 operates. In this system, the operating unit works 80 percent of the time and idles during the remaining 20 percent. Unit 1 has an energized failure rate of λ_{e_1} and a de-energized failure rate of λ_{d_1}. The failure rates for unit 2 and unit 3 are, respectively, λ_{e_2} and λ_{e_3}, λ_{d_2} and λ_{d_3}, and λ_{q_1} and λ_{q_3}, the latter two being failure rates of the standby units prior to operating in the system. (This rate is not the same as the de-energized failure rate.) The sensing and switching devices have failure rates, respectively, of λ_{se_2} and λ_{se_3} while sensing (awaiting switching) and of λ_{sw_2} and λ_{sw_3} while switching. All failure rates are failures/h with the exception of the switching failure rate which is failures/cycle. Determine the reliability of the system.

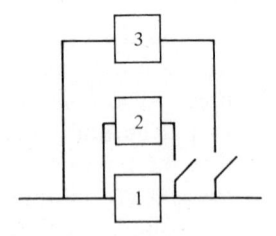

Following the concept of Eq. (4-4), an effective operating failure rate can be found for each unit:

$$\lambda_{op_1} = 0.80\lambda_{e_1} + 0.20\lambda_{d_1}$$
$$\lambda_{op_2} = 0.80\lambda_{e_2} + 0.20\lambda_{d_2}$$
$$\lambda_{op_3} = 0.80\lambda_{e_3} + 0.20\lambda_{d_3}$$

Solution The reliability of the system is:
R = (mode 1) probability that unit 1 functions successfully to time *t* or (mode 2) probability that unit 1 fails at t_1, sense-switch 2 is good at t_1, switch 2 works at t_1, unit 2 is good in quiescent mode to t_1, unit 2 functions properly to *t*.

Or (mode 3) probability that unit 1 fails at t_1, sense-switch 2 is good at t_1, switch 2 works at t_1, unit 2 is good in quiescent mode to t_1, unit 2 fails at t_2, sense-switch 3 is good at t_2, switch 3 works at t_2, unit 3 functions properly to t.

Or (mode 4) probability that unit 1 fails at t_1 sense-switch 2 is good at t_1, switch 2 works at t_1, unit 2 has failed in quiescent mode by t_1, sense-switch 3 is good at t_1, switch 3 works at t_1, unit 3 functions properly to t.

Or (mode 5) probability that unit 1 fails at t_1, sense-switch 2 is good at t_1, switch 2 is bad at t_1, unit 3 functions properly to t.

Or (mode 6) probability that unit 1 fails at t_1, sense-switch 2 fails by t_1, unit 3 functions properly to time t.

For a system of this complexity, it may be helpful to make a table of modes of function, units involved, and time domains. This is done in the following table.

Success mode	1_{op}	2_q	2_{se}	2_{sw}	2_{op}	3_q	3_{se}	3_{sw}	3_{op}
			Units, function, time domain						
1	G $0 \to t$								
2	B t_1	G t_1	G t_1	G 1_c	G $t_1 \to t$				
3	B t_1	G t_1	G t_1	G 1_c	$Gt_1 \to t_2$ Bt_2	G t_2	G t_2	G 1_c	G $t_2 \to t$
4	B t_1	B t_1	G t_1	G 1_c		G t_1	G t_1	G 1_c	G $t_1 \to t$
5	B t_1		G t_1	B t_1		G t_1	G t_1	G 1_c	G $t_1 \to t$
6	B t_1		B t_1			G t_1	G t_1	G 1_c	G $t_1 \to t$

Credit is due Dr. Dimitri Kececioglu of the University of Arizona for the concept of tabular display.

$$R = e^{-\lambda_{op_1} t} + \int_{t_1=0}^{t_1=t} [\lambda_{op_1} e^{-\lambda_{op_1} t_1} e^{-\lambda_{se_2} t_1} e^{-\lambda_{sw_2} \times 1} e^{-\lambda_{q_2} t_1} e^{-\lambda_{op_2}(t-t_1)}] \, dt_1$$

$$+ \int_{t_2=0}^{t_2=t} \int_{t_1=0}^{t_1=t_2} [\lambda_{op_1} e^{-\lambda_{op_1} t_1} e^{-\lambda_{se_2} t_1} e^{-\lambda_{sw_2} \times 1} e^{-\lambda_{q_2} t_2} \lambda_{op_2} e^{-\lambda_{op_2}(t_2-t_1)}]$$
$$\times e^{-\lambda_{se_3} t_2} e^{-\lambda_{sw_3} \times 1} e^{-\lambda_{q_3} t_2} e^{-\lambda_{op_3}(t-t_2)}] \, dt_1 \, dt_2$$

$$+ \int_{t_1=0}^{t_1=t} [\lambda_{op_1} e^{-\lambda_{op_1} t_1} e^{-\lambda_{se_2} t_1} e^{-\lambda_{sw_2} \times 1} \lambda_{q_2} e^{-\lambda_{q_2} t_1} e^{-\lambda_{se_3} t_1} e^{-\lambda_{sw_3} \times 1} e^{-\lambda_{q_3} t_1} e^{-\lambda_{op_3}(t-t_1)}] \, dt_1$$

$$+ \int_{t_1=0}^{t_1=t} [\lambda_{op_1} e^{-\lambda_{op_1} t_1} e^{-\lambda_{se_2} t_1}(1 - e^{-\lambda_{sw_2} \times 1}) e^{-\lambda_{se_3} t_1} e^{-\lambda_{sw_3} \times 1} e^{-\lambda_{q_3} t_1} e^{-\lambda_{op_3}(t-t_1)}] \, dt_1$$

$$+ \int_{t_1=0}^{t_1=t} [\lambda_{op_1} e^{-\lambda_{op_1} t_1} \lambda_{se_2} e^{-\lambda_{se_2} t_1} e^{-\lambda_{se_3} t_1} e^{-\lambda_{sw_3} \times 1} e^{-\lambda_{q_3} t_1} e^{-\lambda_{op_3}(t-t_1)}] \, dt_1 \quad ////$$

PROBLEMS

5-1 Determine the reliabilities of the four simple systems shown below: (operating period in each case is 200 h, unit failure rate is 2.5×10^{-4} failures/h).
  (c) Assume the reliability of the sensing-switching unit is 100 percent.
  (d) Sensing element failure rate is $\lambda_{se} = 3 \times 10^{-5}$ failures/h, switch failure rate is $\lambda_{sw} = 10^{-5}$ failures/cycle.
  (e) Compare the results and draw conclusions. What is the effect of increasing the operating period on the relative reliabilities?

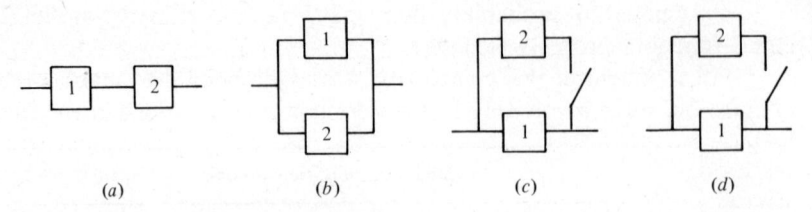

(a)   (b)   (c)   (d)

5-2 (a) Derive the expression for reliability of the system shown if successful operation is defined as requiring a minimum of one unit functioning properly. Assume the sensing-switching unit is a single, composite unit with failure on a per-cycle basis.
  (b) Derive the expression for reliability of the system if successful operation is defined as requiring a minimum of two units functioning properly.

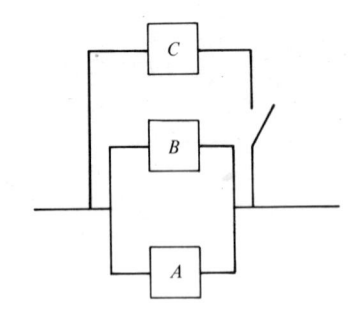

5-3 In the arrangement shown, unit A operates for 55 percent of each operating period (failure rate = λ_{e_A} failures/h) and idles for 45 percent of the period (failure rate = λ_{d_A} failures/h). The sensing device has a failure rate of λ_{se} failures/h. The switch has a failure rate of λ_{sw_q} failures/h (while awaiting switching), a switching failure rate of λ_c failures/cycle, and a fail-open failure rate of λ_{sw_O} failures/h after switching has been accomplished. Unit B has a failure rate of λ_{q_B} failures/h while in standby, an operating failure rate of λ_{e_B} failures/h, and an idling failure rate of λ_{d_B} failures/h.
  (a) Express the reliability of the system in language terms (not in equations).
  (b) Express the reliability of the system in general mathematical terms but do not integrate.
  (c) Integrate the expression derived in part (b) and reduce to a final equation for reliability.
  (d) Simplify part (c) for the case when units A and B are identical.

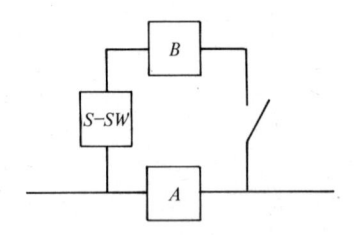

5-4 The system shown requires a minimum of one functioning unit for successful operation. When unit B fails, the failure rate λ_A changes to an increased value, λ_A^*. Likewise, when unit A fails, λ_B changes to an increased value λ_B^*. Assume the sensing-switching unit has a failure rate of λ_{se} failures/cycle. Failure rate of unit C in standby condition is λ_{qc} failures/h and is λ_{ec} failures/h while operating.
  (a) Express the reliability of the system in language terms (not in equations).
  (b) Express the reliability of the system in general mathematical terms. Do not integrate unless desired.

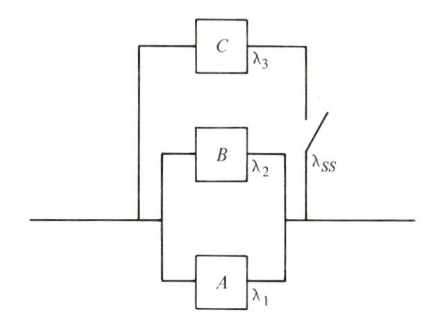

5-5 For the system shown, successful operation requires a minimum of two functioning units. The three units in parallel have failures rates of λ_A, λ_B, and λ_C failures/h, respectively. The standby unit has a failure rate of λ_{q_D} failures/h while in standby and an operating failure rate of λ_{e_D} failures/h. The sensing element has a failure rate of λ_{se} failures/h. The switch has a failure rate of λ_{sw_q} failures/h (while awaiting switching), a switching failure rate of λ_{sw} failures/cycle, and a fail-open failure rate of λ_{sw_0} failures/h after switching.
  (a) Express the reliability of the system in language terms (not in equations).
  (b) Express the reliability of the system in general mathematical terms. Do not integrate unless desired. (A table like that in Example 5-2 may be helpful.)
  (c) Simplify for the case when A, B, and C are identical units.

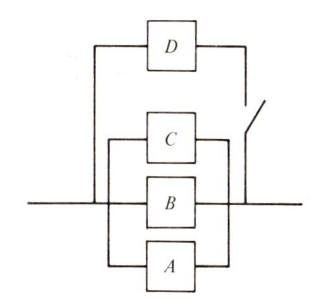

5-6 Rework Prob. 5-5 for the case where system success requires a minimum of three units functioning. Further specialize for the case where units A, B, C, and D are identical.

5-7 Failure rate of a component is independent of its age and equal to 5 percent/1000 h. How many standbys are necessary to ensure a reliability of 99 percent (or more) for 20,000 h of operation if the unreliability of the switching mechanism is negligible? What is the MTTF of the resulting system? (NOTE: Solution is simple but does involve recognition of the proper form to be used.)

5-8 Consider a dc power supply as shown below with a generator ($\lambda_g = 0.0002$ failures/h) and a standby battery with a failure rate of 0.001 failures/h when operating. It is known that the sensing-switching device has a 99 percent reliability for a single switching operation. Assume the failure rate of the battery is zero when it is in standby. Derive the reliability equation for a mission time of t h. What is the numerical value of reliability for $t = 10$ h? Interpret the results.

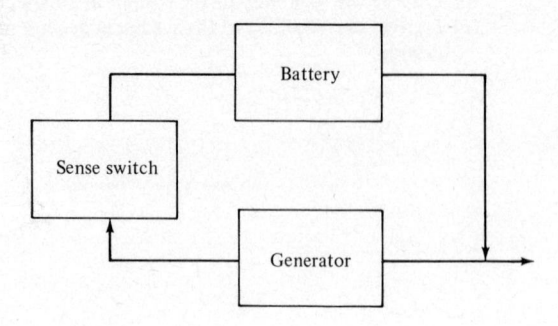

6
CONDITIONAL PROBABILITY

6-1 INTRODUCTION

There are problems in calculating reliability in which the system configuration cannot be reduced to simple series, parallel, or standby models, nor to obvious mixtures of these. There are some systems in which the combinations are none of these, in the usual sense. There may also be combinations when one is simply not sure enough to determine such a model. There are other situations in which a component (or system) is expected to perform a dual function. In both of these general situations, reliability calculations can be aided by using conditional probabilities. (See Appendix A.)

6-2 DUAL FUNCTION

The systems considered in Chaps. 4 and 5 are essentially *control* types, i.e., a signal is supplied at specified times and not supplied at others. The circuit can be open when it should be closed (essentially the case considered) or the circuit can be closed when it should be open (a premature failure). The basic equations developed earlier are still valid. It is also appropriate, however, to use conditional probabilities. These must be properly interpreted.

64 INTRODUCTION TO RELIABILITY IN DESIGN

If the condition of interest is an open circuit, then R is the conditional probability of success (reliability) and Q is the conditional probability of a closed circuit, i.e., premature closure (unreliability).

If the condition of interest is a closed circuit, R is the conditional probability of success (reliability) and Q is the conditional probability of an open circuit (unreliability).

EXAMPLE 6-1 Consider the system (shown schematically in the figure) in which a battery is connected to an electric match which fires a warhead on a ballistic missile. The circuit between the battery and match is completed when the acceleration switch senses enough g load to close and one (or both) of the radars

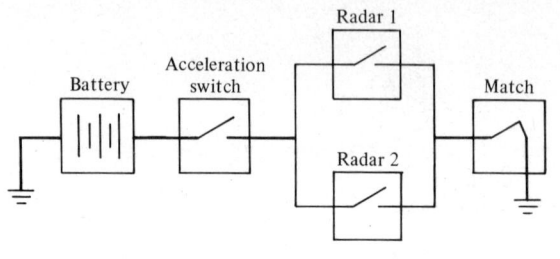

detects a target and closes a switch. In the absence of target and g load, an explosion is not wanted (condition 1). When a target and g load are present, firing is desired (condition 2). Calculate the reliability and unreliability for each condition. Assume independent units.

Condition 1
$R_1 = P(\text{no explosion})$
$Q_1 = P(\text{premature explosion})$
$P_{B_1} = P(\text{battery, good})$
$P_{A_1} = P(\text{acceleration switch, closes})$
$P_{R_{11}} = P(\text{radar 1 switch, closes})$
$P_{R_{21}} = P(\text{radar 2 switch, closes})$
$P_{M_1} = P(\text{match, good})$
$P_{H_1} = P(\text{warhead, ignites})$
$Q_1 = P_{B_1} P_{A_1} P_{M_1} P_{H_1} (P_{R_{11}} + P_{R_{21}} - P_{R_{11}} P_{R_{21}})$
$R_1 = 1 - Q_1$

Condition 2
$R_2 = P(\text{success})$
$Q_2 = P(\text{dud})$
$P_{B_2} = P(\text{battery, good})$
$P_{A_2} = P(\text{acceleration switch, closes})$
$P_{R_{12}} = P(\text{radar 1 switch, closes})$
$P_{R_{22}} = P(\text{radar 2 switch, closes})$
$P_{M_2} = P(\text{match, good})$
$P_{H_2} = P(\text{warhead, ignites})$
$R_2 = P_{B_2} P_{A_2} P_{M_2} P_{H_2} (P_{R_{12}} + P_{R_{22}} - P_{R_{12}} P_{R_{22}})$
$Q_2 = 1 - R_2$

The equations for Q_1 and R_2 appear to be identical except for the difference in subscripts between condition 1 and condition 2. The probabilities of a good battery, a good electric match, and ignition of the warhead should be the same in both conditions. The probabilities of the acceleration switch and the radar switches closing should differ greatly between the two conditions. If typical probability values are:

$$P_{R_{11}} = P_{R_{21}} = 0.01 \qquad P_{B_1} = P_{B_2} = 0.98$$
$$P_{A_1} = 0.001 \qquad P_{M_1} = P_{M_2} = 0.99$$
$$P_{R_{12}} = P_{R_{22}} = P_{A_2} = 0.99 \qquad P_{H_1} = P_{H_2} = 0.99$$

then

$$Q_1 = (0.98)(0.001)(0.99)(0.99)(0.01 + 0.01 - 0.0001) = 0.0000191$$
$$R_1 = 0.9999819$$
$$R_2 = (0.98)(0.99)(0.99)(0.99)(0.99 + 0.99 - 0.9801) = 0.950798$$
$$Q_2 = 0.049202 \qquad ////$$

In a number of situations, the operating conditions of the system may be clear but there may be a need for conditional probabilities of some components within the system. There are other circumstances in which more than two events must be defined for one component (or more).

EXAMPLE 6-2 Consider a simplified system to control the power delivered to system A in the sketch below. What is the reliability for reception of power at A, given (1) the control switch is on, and (2) A does not draw an overload? For

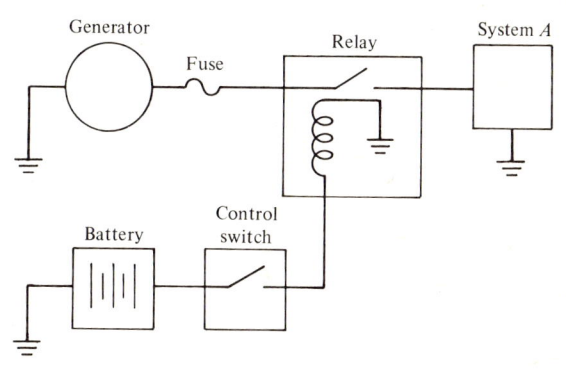

independent units, the reliability for reception of power at A is the probability that the generator is good and that the fuse and relay provide a closed circuit, thus

$$R = P_G P_F P_C$$
$$P_G = P(\text{generator, good})$$
$$P_F = P(\text{fuse, good})$$
$$P_C = P(\text{relay contact, closed})$$

A fuse has two functions: to pass subnormal and normal currents and to open under excessive current. We were given a no-overload condition. The probability P_F is therefore actually a conditional probability, but is used as a complete probability.

The relay contact can be closed either because the relay receives power (proper operation) or because the contact fails when it is closed with no power. The probability of the contact being closed is [Eq. (A-7)]

$$P_c = P(C_{R/E})P(E) + P(C_{R/\bar{E}})P(\bar{E})$$

where C_R is closed relay contact and E is the power to the relay. $P(E)$ can be estimated as

$$P(E) = p_b p_s$$

where p_b = probability of a good battery and p_s = probability of a good control switch, given it is turned on. The system reliability becomes

$$R = p_g p_f [P(C_{R/E}) p_b p_s + P(C_{R/\bar{E}})(1 - p_b p_s)]$$

It should be noted that it is feasible to run tests to determine values for all terms in this equation. We would also expect that $P(C_{R/E})$ would be close to unity while $P(C_{R/\bar{E}})$ would be close to zero. ////

6-3 BAYES' THEOREM

For situations where uncertainty exists as to the reliability configuration of the system, use of Bayes' theorem [Eq. (A-9)] can be especially helpful. Caution must be used, however, in applying it. Bayes' theorem essentially combines the multiplication law of probability with the law of total probability to give an expression for a posteriori probabilities which involve certain a priori and conditional probabilities associated with specified events.

In the block diagram shown in Fig. 6-1, two parallel paths XX' and YY' operate to assure system output if at least one of the paths is good. Since neither X nor Y is sufficiently reliable, a third element Z is added to supply either X' or Y'. This means that four paths are possible: XX', YY', ZX', and ZY'. (Try calculating the system reliability.) If the system were connected as shown in Fig. 6-2 instead of using the connections shown in Fig. 6-1, there

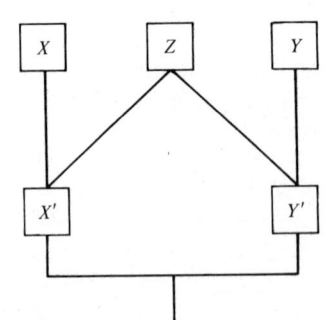

FIGURE 6-1
Reliability block diagram of system which is neither series, parallel, nor standby.

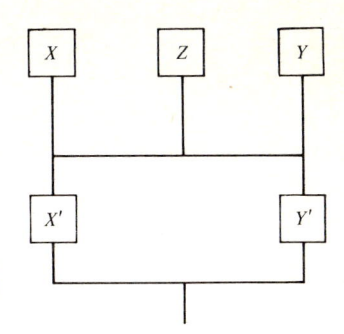

FIGURE 6-2
Reliability block diagram of system in Fig. 6-1 modified to two parallel subsystems which are in series.

would be three elements, X, Y, and Z, in parallel in the upper row. This trio is in series with the lower row of two elements, X' and Y', in parallel. System reliability is easy to calculate in this case. System reliability for the arrangement in Fig. 6-1 can be calculated using Bayes' theorem. System reliability in Fig. 6-2 is $[1 - (1 - R_X)(1 - R_Y)(1 - R_Z)][1 - (1 - R_{X'})(1 - R_{Y'})]$ while system reliability in Fig. 6-1 is

$$1 - R_Z(1 - R_{X'})(1 - R_{Y'}) - (1 - R_Z)(1 - R_X R_{X'})(1 - R_Y R_{Y'}).$$

EXAMPLE 6-3   A guided missile M is equipped with three guidance systems A, B, and C. A is highly accurate. It is complex, and test experience has shown it to have a reliability of 90 percent. B is less complex and is less accurate. It is more reliable, at 95 percent. C is relatively crude with low accuracy but with a reliability of 98 percent. The missile is fired. It misses the target and disappears. Flight data indicate mission failure was almost certainly due to failure in guidance. The probability of more than one guidance system failing in flight is negligible. What is the probability of failure due to each guidance subsystem? Denote missile failure by M_F and guidance system failures by A_F, B_F, and C_F respectively. Careful system analysis has shown that $M_F/A_F = 0.20$, $M_F/B_F = 0.05$, and $M_F/C_F = 0.01$.

$$P\left(\frac{A_F}{M_F}\right) = \frac{0.20 \times 0.10}{(0.20)(0.10) + (0.05)(0.05) + (0.01)(0.02)} = 0.881$$

$$P\left(\frac{B_F}{M_F}\right) = 0.110$$

$$P\left(\frac{C_F}{M_F}\right) = 0.009$$

For this problem, failure of the missile has a high probability (nearly 90 percent) of being caused by failure of guidance system A. The redundancy designed into the system is largely ineffective. It would appear that B and C should be eliminated and effort devoted to making A more reliable.

(This problem was fabricated to make a point. It is not a practical application.)

////

68 INTRODUCTION TO RELIABILITY IN DESIGN

As indicated above, Bayes' theorem is an a posteriori probability which requires use of a priori and conditional probabilities. The use of Bayes' theorem can be enlarged from the original intent to give the reliability of a system as follows: The reliability of a system is equal to the reliability of the system, given that an arbitrarily selected unit is good, multiplied by the reliability of that unit *plus* the reliability of the system, given that the selected unit is bad, multiplied by the unreliability of that unit. In mathematical form.

$$R_{\text{sys}} = R(\text{sys}/A_{\text{good}})R(A) + R(\text{sys}/A_{\text{bad}})Q(A) \qquad (6\text{-}1)$$

In the same manner, the system unreliability can be expressed as

$$Q_{\text{sys}} = Q(\text{sys}/A_{\text{good}})R(A) + Q(\text{sys}/A_{\text{bad}})Q(A) \qquad (6\text{-}2)$$

These equations can be applied to all combinations of units in a given system. They are particularly useful when it is (or appears to be) difficult to determine that the system, as shown by the reliability block diagram, is clearly series, parallel, or standby. It should be noted that there are many situations in which the arrangement is not clearly or readily reducible to simple systems. If the reliability block diagram shows that the system is clearly series, parallel, or standby, it may not be expedient to use Eqs. (6-1) or (6-2). If there is any doubt, one can use Bayes' theorem, either directly or as a check on some other procedure.

If it is not possible to write terms such as $R(\text{sys}/A_{\text{good}})$, $R(\text{sys}/A_{\text{bad}})$, etc., directly, they can be obtained by repeated use of Eqs. (6-1) and/or (6-2) until each term can be written directly in terms of the reliabilities and unreliabilities of all the individual units in the system.

EXAMPLE 6-4 Find the reliability of the system shown using Bayes' theorem when system success requires that at least one of the following paths is good: *AB*, *BC*, or *DE*.

With C_{good} system is

$$R(\text{sys}/C_{\text{good}}) = 1 - (1 - R_D R_E)(1 - R_B)$$

With C_{bad}, system is

$$R(\text{sys}/C_{\text{bad}}) = 1 - (1 - R_A R_B)(1 - R_D R_E)$$

$$R_{\text{sys}} = [1 - (1 - R_D R_E)(1 - R_B)]R_C$$
$$\quad + [1 - (1 - R_A R_B)(1 - R_D R_E)](1 - R_C)$$

$$R_{\text{sys}} = R_A R_B + R_B R_C + R_D R_E - R_A R_B R_C - R_A R_B R_D R_E$$
$$\quad - R_B R_C R_D R_E + R_A R_B R_C R_D R_E$$

CONDITIONAL PROBABILITY   69

With C good, system is

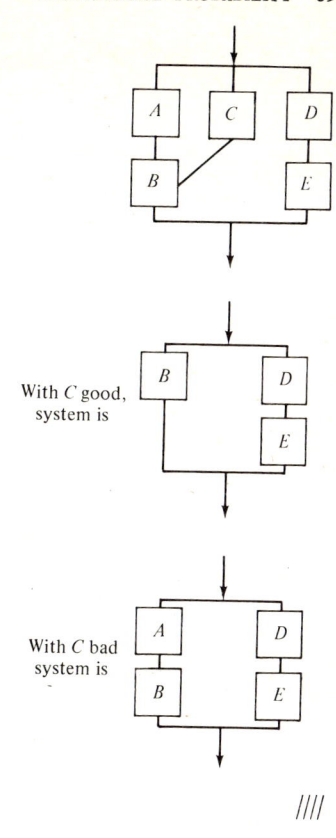

With C bad system is

EXAMPLE 6-5   Use Bayes' theorem to determine system reliability when three out of four units are required for system success.

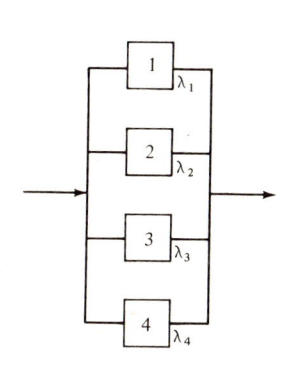

$$R_{\text{sys}} = R(\text{sys}/\#1_{\text{good}})R_1 + R(\text{sys}/\#1_{\text{bad}})Q_1$$
$$= R[(\text{sys}/\#2_{\text{good}})R_2 + (\text{sys}/\#2_{\text{bad}})Q_2]R_1 + R(\text{sys}/\#1_{\text{bad}})Q_1$$
$$= \{[1 - (1 - R_3)(1 - R_4)]R_2 + R_3R_4(1 - R_2)\}R_1 + R_2R_3R_4(1 - Q_1)$$
$$= R_1R_2R_3 + R_1R_2R_4 + R_1R_3R_4 + R_2R_3R_4 - 3R_1R_2R_3R_4$$

From binomial theorem

$$(R_1 + Q_1)(R_2 + Q_2)(R_3 + Q_3)(R_4 + Q_4) = 1$$

$$\begin{aligned}R_{\text{sys}} &= R_1R_2R_3R_4 + R_2R_3R_4Q_1 + R_1R_3R_4Q_2 + R_1R_2R_4Q_3 + R_1R_2R_3Q_4 \\ &= R_1R_2R_3 + R_1R_2R_4 + R_1R_3R_4 + R_2R_3R_4 - 3R_1R_2R_3R_4\end{aligned} \quad ////$$

PROBLEMS

6-1 Given the circuit shown with the following probabilities:
$p_B = P(\text{battery, good})$
$p_{Ag} = P(\text{acceleration switch is closed after actual acceleration})$
$p_{Ab} = P(\text{acceleration switch is closed before actual acceleration})$
$p_{o1} = P(\text{resistor 1 is open})$
$p_{o2} = P(\text{resistor 2 is open})$
$p_{s1} = P(\text{resistor 1 is grounded})$
$p_{s2} = P(\text{resistor 2 is grounded})$
$p_{g1} = P(\text{resistor 1 is good})$
$p_{g2} = P(\text{resistor 2 is good})$
$p_L = P(\text{light, good})$
Assume:
$p_{o1} + p_{s1} + p_{g1} = 1$
$p_{o2} + p_{s2} + p_{g2} = 1$

Assuming the subexperiments are independent, find the reliability for (*a*) light operating after actual acceleration, and (*b*) light not operating before actual acceleration.

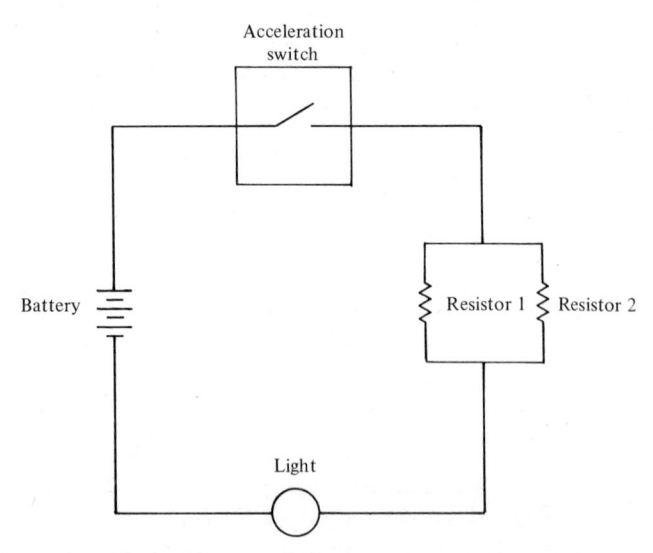

6-2 Given the circuit shown with the defined probabilities, find the reliability for (*a*) the light operating after the specified time, and (*b*) the light operating before the specified time.
$p_B = P(\text{battery, good})$
$p_{Xb} = P(\text{timer } X, \text{ closed before specified time})$
$p_{Xa} = P(\text{timer } X, \text{ closed after specified time})$

$p_{Yb} = P$(timer Y, closed before specified time)
$p_{Ya} = P$(timer Y, closed after specified time)
$p_L = P$(light, good)

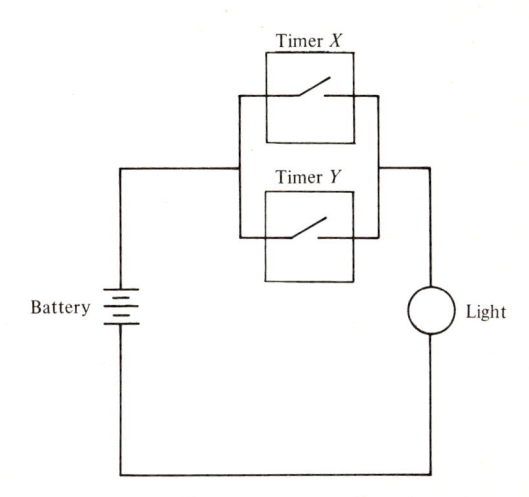

6-3 A system is composed of three subsystems K, L, and M. All three must work in order to have system success. There are two environments E_1 and E_2 in which the system can work. Given
$P(KLM/E_i) = P(K/E_i)P(L/E_i)P(M/E_i)$ for $i = 1, 2$
$P(K/E_1) = P(L/E_1) = P(M/E_1) = 0.9$
$P(K/E_2) = 0.9 \quad P(L/E_2) = 0.9 \quad P(M/E_2) = 0.7$
The two environments are equally likely.
(a) Find the reliability of the system, given environment E_1.
(b) Find the reliability of the system.
(c) It is proposed that an identical system be added in parallel redundancy (one functioning required for success). Find the reliability of the new system. Are the answers different if we assume conditional independence in comparison with an assumption of independence? If they are different, why? Which assumption is more reasonable, if there is a difference?

6-4 Verify the reliabilities of the systems shown in Figs. 6-1 and 6-2.

6-5 The system shown below was worked in Example 6-4 using unit C as the "pivot" unit. Rework the problem using unit B as the pivot unit. Success requires that at least one of the paths AB, BC, or DE is good.

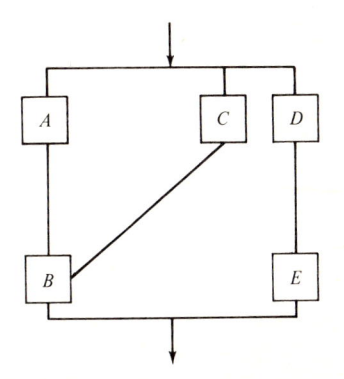

6-6 Given the system shown, find its reliability if system success requires that at least one of the following paths be good: *ABG, ACG, DCG, DEF*. (If Bayes' theorem is used, it may be necessary to use it more than once or twice.)

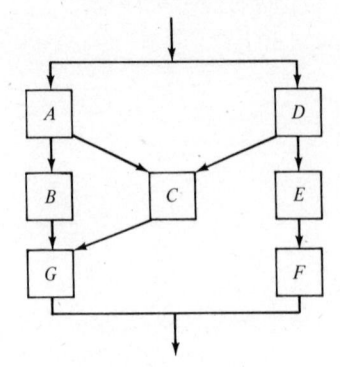

6-7 Given the following schematic of a system, use Bayes' theorem to find its reliability if system success requires that at least one of the following paths be good: *ABG, DBG, DCG, DEF*.

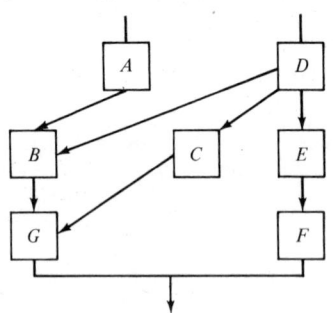

6-8 A binary communication channel carries data as one of two types of signals, i.e., ones or zeros A transmitted zero is sometimes received as a one and a transmitted one is sometimes received as a zero because of noise. For a given channel, assume a probability of 0.94 that a received zero is a transmitted zero and a probability of 0.91 that a received one is a transmitted one. Further assume a probability of 0.45 of transmitting a zero. If a single signal is sent, determine
(a) probability that a one is received,
(b) probability that a zero is received, and
(c) probability that a one was transmitted if a one was received.

6-9 A given lot of small devices is 98 percent good and 2 percent defective. To be certain of using a good device, each device is tested before installation. The tester itself if not totally reliable since it has the following conditional probabilities:
P(says good/actually good) = 0.95
P(says bad/actually bad) = 0.95
A device is tested with the tester indicating the device is bad. What is the probability the device is actually bad? Before calculating the answer, make an educated guess.

6-10 Three boxes each contain two units. Both units in the first box are good. One unit in the second box is good while the second is bad. Both units in the third box are bad. A technician selects both a box and a unit from it at random. A test check shows that the unit is good.
(a) What is the probability that the other unit in the same box is also good?
(b) What is the probability that the other unit in the same box is bad?

6-11 A twin-engine aircraft needs an electrical generating system. Three possible systems are shown below in block diagram form. Only the major components are indicated: engines, generators, and frequency changers. The generators are direct, engine-driven, variable-frequency changers. The static-frequency changers convert the generator output into constant-frequency power. The frequency changers in *A* are rated at 60 kVA, but they are rated at 30 kVA in the other two configurations. In *C*, the frequency changer in the middle can be switched automatically or manually to either of the two generators shown.

Successful operation is defined as the system supplying 60 kVA of normal power and 30 kVA of emergency power.

Find the reliabilities (in general terms). Take useful life failure rates to be: engines—0.00005 failure/h; generators—0.0001 failure/h; frequency changers—0.0005 failure/h. Compare the reliabilities of the three systems and indicate which system you would use (all other factors being equal).

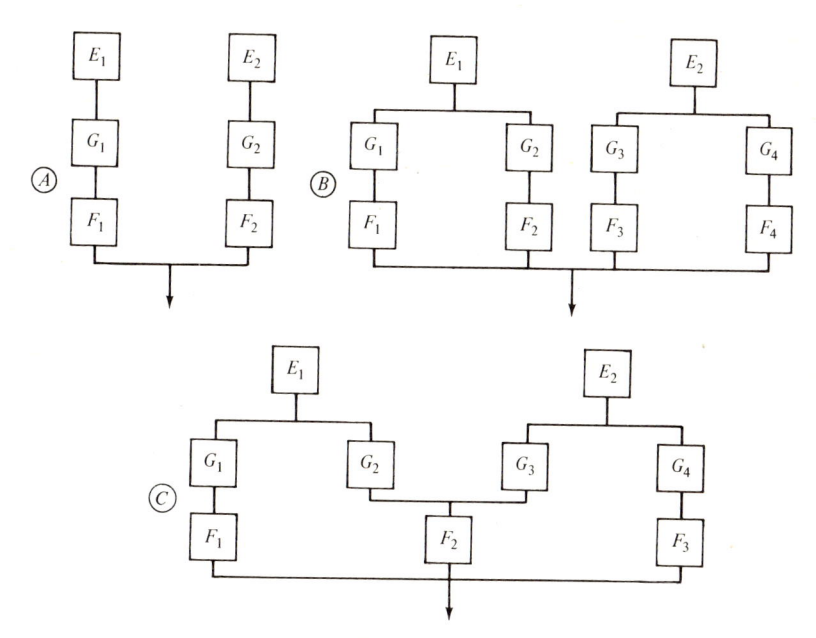

7[†]
MULTIMODE FUNCTION

7-1 INTRODUCTION

In the preceding chapters, discussion has implied that most components or devices either perform properly or malfunction. While this is true, it is an oversimplification in many situations, since there may be a number of types of malfunction, some of which are more serious than others. The ability to differentiate between various kinds of malfunctions allows a more precise definition and calculation of reliability. A switch or valve, for example, may function properly, but it can fail in a number of ways; e.g., it may fail open when it is expected to shut, it may fail shut when it is expected to open, it may fail open before it is expected to open, it may fail shut before it is expected to shut, etc. In any system with several devices, many of which can function in various ways, the number of terms which must be evaluated to obtain system reliability can become rather large. In a situation of this sort, it is highly desirable to proceed systematically to avoid missing any terms.

[†] Credit is due to Dr. Dimitri Kececioglu for the basic concept presented in this chapter.

7-2 ONE POSSIBLE PROCEDURE

1 Carefully and completely define system success.

2 Calculate the total number of permutations (Sec. A-6), with repetitions permitted, of all possible ways of combining all units in the system in all possible individual modes of functioning. The number of permutations is

$$\text{Perm} = M_1^{N_1} \times M_2^{N_2} \times M_3^{N_3} \times \cdots \times M_n^{N_n} \qquad (7\text{-}1)$$

where M_i = number of modes of function of device i in the system
N_i = number of devices in the system having M_i modes of function

Consider, for example, a system composed of five units, one unit (N_1) having two modes of function (M_1), two units (N_2) having four modes of function (M_2), and the remaining two units having three modes of function (M_3). The number of permutations is $2^1 \times 4^2 \times 3^2 = 288$.

3 Using any systematic procedure of your own devising, determine all the possible permutations of the modes of function of all units in the system, making sure that all permutations are included and that none are duplicated.

4 System output should be calculated for each permutation and should then be measured against the criteria for system success. Keep a record of success or nonsuccess from each permutation.

5 Calculate the probability of system success for each permutation in terms of appropriate probabilities of individual units. This is commonly done only for those permutations which give system success (although the probability of nonsuccess can be calculated for the other permutations). In most circumstances, the various probability terms are independent of each other. If this is not the case, then appropriate conditional probabilities must be used.

6 System reliability is the sum of probabilities from all the permutations which give system success. If the probability of each permutation, whether a system success or not (step 5), is calculated, the sum of the probabilities of all permutations should be equal to 1. This can be used as a check on the work.

EXAMPLE 7-1 Consider a set of three diodes in parallel as shown. Each diode can function properly but can malfunction by short-circuiting or by open-circuiting.

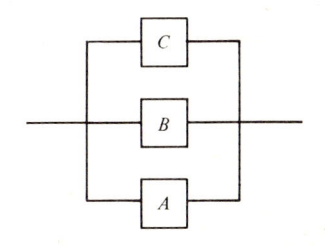

Let the probabilities of these be

p_n = probability of proper operation
p_s = probability of short-circuiting
p_o = probability of open-circuiting

System success is defined as flow of a controlled quantity of electrical current. Assume diodes and failure modes are independent, i.e., $p_n + p_s + p_o = 1$. Determine the system reliability.

From Eq. (7-1):

$$\text{Perm} = 3^3 = 27$$

The following table is one type of systematic procedure.

Permutations	Units: modes of function			System success	Probability of system success
	A	B	C		
1	n	n	n	yes	$p_{nA} p_{nB} p_{nC}$
2	n	n	s	no	—
3	n	n	o	yes	$p_{nA} p_{nB} p_{oC}$
4	n	s	n	no	—
5	n	s	s	no	—
6	n	s	o	no	—
7	n	o	n	yes	$p_{nA} p_{oB} p_{nC}$
8	n	o	s	no	—
9	n	o	o	yes	$p_{nA} p_{oB} p_{oC}$
10	s	n	n	no	—
11	s	n	s	no	—
12	s	n	o	no	—
13	s	s	n	no	—
14	s	s	s	no	—
15	s	s	o	no	—
16	s	o	n	no	—
17	s	o	s	no	—
18	s	o	o	no	—
19	o	n	n	yes	$p_{oA} p_{nB} p_{nC}$
20	o	n	s	no	—
21	o	n	o	yes	$p_{oA} p_{nB} p_{oC}$
22	o	s	n	no	—
23	o	s	s	no	—
24	o	s	o	no	—
25	o	o	n	yes	$p_{oA} p_{oB} p_{nC}$
26	o	o	s	no	—
27	o	o	o	no	—

System reliability is

$$R = p_{nA}p_{nB}p_{nC} + p_{nA}p_{nB}p_{oC} + p_{nA}p_{oB}p_{nC} + p_{oA}p_{nB}p_{nC} + p_{nA}p_{oB}p_{oC}$$
$$+ p_{oA}p_{nB}p_{oC} + p_{oA}p_{oB}p_{nC}$$

If all three diodes are identical, then system reliability is

$$R = p_n^3 + 3p_n^2 p_o + 3p_n p_o^2$$

////

PROBLEMS

7-1 Rework Example 7-1 with
   (a) one unit as given, but the other two having only two modes of function, i.e., normal and failing short-circuited.
   (b) two units as given, but the third having only two modes of function, i.e., normal and failing short-circuited.
   (c) one unit as given, but the other two having only two modes of function, i.e., normal and failing open-circuited.
   (d) two units as given but the third having only two modes of function, i.e., normal and failing open-circuited.

7-2 A radioactive liquid flows continuously through a nuclear reactor system. To shut off flow, two valves are physically in series as shown. The two valves are identical. Failure rates and modes are:

λ_{od} = fail open while de-energized
λ_{oe} = fail open when energized (signaled to shut)
λ_{sd} = fail shut while de-energized and stay shut when signaled to open
λ_{se} = fail shut when energized (signaled to shut) and stay shut, when signaled to open
λ_{∞} = fail open after shutting but before being signaled to open

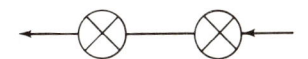

   (a) It is most important that there be no stoppage of flow unless shutoff is intended, but flow must stop when shutoff is intended and stay stopped until signaled to resume. Resumption of flow upon signal is not important.
   Determine the subsystem reliability (use R and Q).
   (b) There must be no stoppage of flow unless shutoff is intended, but flow must stop when shutoff is intended and stay stopped until signaled to resume. Flow must resume upon command or signal to resume. Determine subsystem reliability (use R and Q).
   (c) Rework parts (a) and (b) if only three modes of function are considered, i.e., normal, failed open, and failed shut. (This can be done by rationalization without rederiving.)
   (d) Rework parts (a) and (b) if only two modes of function are considered, i.e., normal and failed. [Rationalize as in part (c).]
   (e) Compare the results and draw conclusions.

7-3 This problem is identical with Prob. 7-2 except that the two valves are physically in parallel as shown. Add part (f): Compare the results with those of Prob. 7-2 and draw conclusions.

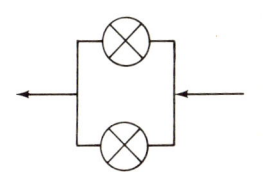

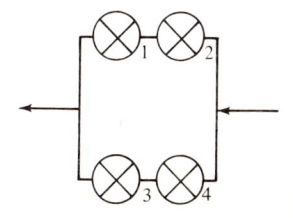

7-4 (a) Consider a system similar to that in Prob. 7-2 but with two valves physically in series in each of two lines which, in turn, are parallel as shown. Determine the reliability of the valve system if success is defined as stopping flow when stoppage of flow is intended. Consider that the valves function normally, have failed open, or have failed shut. Are the physical and reliability block diagrams identical? Why?

    (b) Rewrite if there are only two modes of operation, i.e., normal function and failed.

7-5 Three unequal switching devices are physically connected in series. All three can function normally, fail open, or fail shut.

    (a) What is the system reliability if current is normally flowing and system success is defined as stopping flow on command?

    (b) If these devices are actually regulators whose function is to control the quantity of flow on a continuous basis, what is the reliability of the system?

    (c) Is it possible to get different answers using the same approach? If so, how?

    (d) Is it possible to get the same answers using different approaches? If so, how?

8
DERATING AND MAINTENANCE

8-1 DERATING

Components and devices are commonly designed to sustain certain nominal stresses in operation. Time, through aging effects, is only one such stress. Others are such items as heat, humidity, corrosion, mechanical stresses (direct loads, vibration, shock, etc.), electrical stresses (voltage, current, frequency, etc.), and so on. When a population of devices is operated under the rated conditions, a certain failure rate is observed. This is known as a nominal, basic, or generic failure rate.

It is well known that, as operational stresses increase above the rated level, the observed failure rate increases above the nominal failure rate. Likewise, the observed failure rate decreases when stresses are decreased below the rated level. An indication of this effect is given in Sec. 5-1. The general effects on the mortality curve are shown in Fig. 8-1.

In principle, all the stresses could be described by a stress vector $S\{t, x_1, x_2 \ldots, x_i, \ldots, x_n\}$ where the x_i signify the various stresses. In principle, the failure rate of a device could be measured under a wide variety of conditions S so that a failure-rate vector $\lambda\{t, x_1, x_2, \ldots, x_i, \ldots x_n\}$ could be constructed. This might be in the form of equations or curves. This

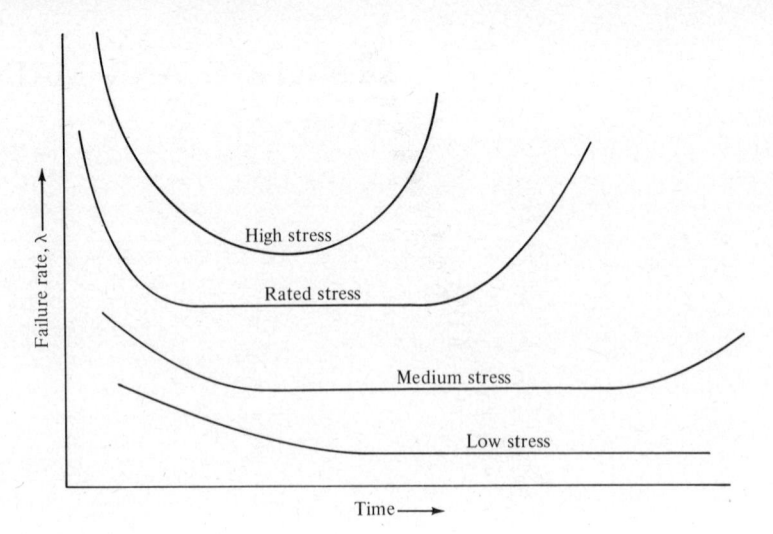

FIGURE 8-1
Effect of environmental stresses on failure rate, schematic.

vector indicates interdependence of stresses. This occurs in many situations; e.g., electrical conductivity varies with temperature. In addition to a large number of such known interdependencies, there are others which are not readily formulated; e.g., a device will fail if the temperature exceeds a critical value, but the device may continue to operate if the critical temperature is exceeded only very briefly. Some devices may be able to withstand high temperatures but not rapid changes of temperature; i.e.; they are susceptible to thermal shock. Some stresses, such as corrosion, are difficult to quantify or to measure readily. The stresses to which a device is subjected during operating life are random. Therefore, the stress vector generally will be a multidimensional, nonstationary, random process. Solution, in the general form, is not easy. Solutions to most problems involving environmental stresses are lacking.

The highest value of any given type of stress that a device can sustain is called its strength. If the applied stress exceeds the strength, the device will fail. For example, the dielectric strength of an insulator is the voltage above which the insulator breaks down and electrical discharge occurs. The rated strength is, of course, the measure of central tendency (Sec. A-9), i.e., a mean of some kind, of a distribution of values. If we have a large number (N) of devices at temperature, humidity, or other stress (x_i), then $Np(x_i) \, dx_i$ devices will fail in the stress interval dx_i. In other words, the probability of failure of any one device near stress level x_i is $p(x_i) \, dx_i$. The stress to which the device is subjected (not in strength testing, but in actual operation) is also a random variable. (It might be more accurately called a random process.) Its distribution is called the stress distribution. Failure of the device will occur in the region where the two distributions overlap, as indicated in Fig. 8-2.

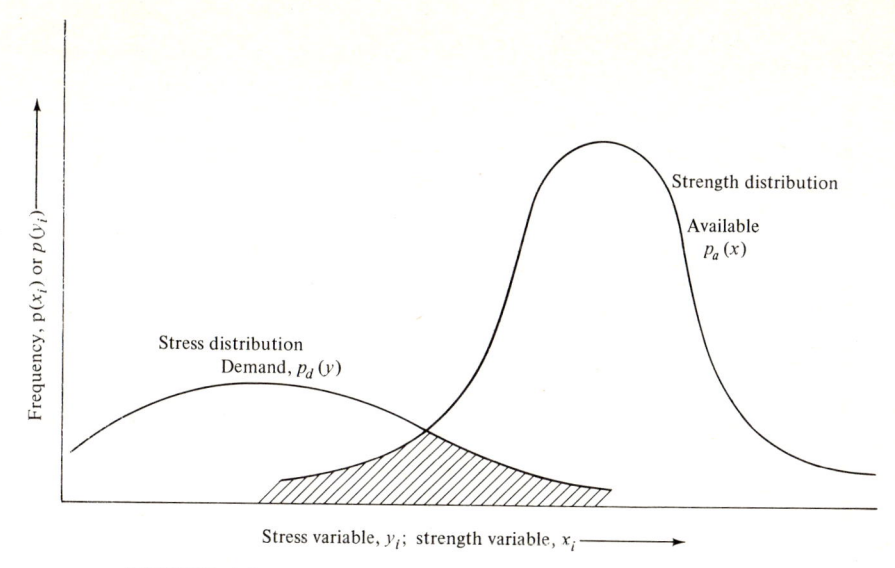

FIGURE 8-2
Distribution of stress demand and available strength, schematic.

Let the stress variable (demand) have a probability density function of $p_d(y)$ and the strength variable (strength available or capacity) have a probability density function of $p_a(x)$. It is obvious that stress and strength are independent. Therefore, the reliability [probability of available strength (X) exceeding demand (Y)] is

$$R = P(X > Y) = \int_0^\infty \left[\int_y^\infty p_a(x)\, dx \right] p_d(y)\, dy \\ = \int_0^\infty \left[\int_0^x p_d(y)\, dy \right] p_a(x)\, dx \tag{8-1}$$

Determination of reliability of a device under stress can be approached through the failure rate. Let the stress variable (e.g., temperature) be θ. Until now, we have implicitly assumed θ to be constant but the reliability $R(t)$ in reality was $R(t/\theta)$. If the probability density function of θ is $p(\theta)$, then, from the theorem of total probability

$$R(t) = \int_{\theta_1}^\infty R\left(\frac{t}{\theta}\right) p(\theta)\, d\theta \tag{8-2}$$

By testing devices at a given stress level, we can find the failure rate $\lambda(t/\theta)$. If this is repeated for various levels of θ,† we can find the function $\lambda = f(t,\theta)$ from which

$$R\left(\frac{t}{\theta}\right) = \exp\left[-\int_0^t f(\tau,\theta)\, d\tau \right] = e^{-\Lambda(t,\theta)} \tag{8-3}$$

† Factorial experiments may be very useful in this context.

Substitution in Eq. (8-2) gives

$$R(t) = \int_{\theta_1}^{\infty} e^{-\Lambda(t,\theta)} p(\theta)\, d\theta \qquad (8\text{-}4)$$

Evaluation of Eq. (8-2) or Eq. (8-4) is seldom simple, and in many cases it is essentially impossible with the data available.

Even though Eq. (8-4) may be impossible to evaluate, devices do fail with failure rates and reliabilities which depend on imposed stress levels. A reduction in stress level, as indicated, reduces the failure rate. In many design situations, however, stress levels are fixed. The same relative effect of reduced stress level can be attained by selecting a device which is rated in excess of the design requirements. This practice of using components or devices at stress levels below their rated stress levels is known as *derating*. Through derating, a larger safety margin is established for peak stresses or for long periods of operation.

Derating can be accomplished in a variety of ways. A crude assessment of the effect of temperature can be made using the Arrhenius equation

$$\text{Rate} = Ae^{-Q/RT} \qquad (8\text{-}5)$$

where A is a "constant," Q is the activation energy, R is the universal gas constant, and T is the absolute temperature. In many chemical reactions, for example, the reaction rate approximately doubles for a 10°C increase in temperature. Behavior of a number of electrical components, e.g., capacitors, correlates reasonably well with this equation.

In some components, the life is approximately inversely proportional to the fifth power of the stresses. An electrical device following this "law," if operated at 1.2 times its rated voltage, would have its life reduced to approximately one-third the life at rated voltage. The same component, operated at 80 percent of rating, would have its life approximately tripled. The same component, operated at 50 percent of rating, would have its life increased by a factor of about 30 over the rated life.

One rather general approach follows from the concept that the failure rate of a device is determined by the various environmental and operational stresses imposed on the device. Each stress is responsible for a portion of the observed failures, i.e., in effect, a partial failure rate can be correlated with each of the stresses. The total failure rate under the simultaneous influence of these stresses is the sum of the partial failure rates:

$$\lambda = \lambda_1 + \lambda_2 + \lambda_3 + \cdots + \lambda_a + \lambda_b + \lambda_c + \cdots \qquad (8\text{-}6)$$

where $\lambda_1, \lambda_2, \lambda_3, \ldots$, are governed by the environmental stresses and $\lambda_a, \lambda_b, \lambda_c, \ldots$, are governed by the operational stresses. Environmental stresses are those which are present whether or not the device is in active operation. They include temperature, humidity, pressure, radiation, ambient chemicals, microbes, etc. Operating stresses appear only when the device is in active operation. They include mechanical loading, voltage, current, frequency, self-generated heat, etc. Such stresses as vibration, shock, and acceleration can fall in either category, depending on the context of the application. Obviously, it requires an extensive stress analysis and data which are suitable to allow calculation of Eq. (8-6).

A somewhat similar approach uses a series of correction factors. An extensive stress analysis is necessary. This is useful in focusing on potential areas of unreliability, even though the predicted reliability may be less accurate than desired.

Correction factors are applied in the form

$$\lambda = \lambda_0(K_1 K_2 K_3 K_4 K_5 K_6) \qquad (8\text{-}7)$$

where λ = the adjusted ("corrected") failure rate
λ_0 = the nominal (generic) failure rate
K_1 = a correction for applied stresses
K_2 = the ratio of expected failures within tolerance to random catastrophic failures
K_3 = an adjustment for changes in environmental stresses
K_4 = an adjustment for different maintenance practices which can affect system failures
K_5 = a correction for system complexity; the more complex the system, the greater the failure rate
K_6 = an adjustment for cycling effects

Equation (8-7) is sometimes "simplified" by grouping all correction factors into two: K_{ap} (including input-output, fraction of rating, etc., i.e., operational stresses) and K_{en} (other categories, i.e., environmental stresses). Equation (8-7) then becomes

$$\lambda = \lambda_0 K_{ap} K_{en} \qquad (8\text{-}8)$$

It is obvious that extensive data are necessary to determine these correction factors.

Data of various kinds are available from a variety of sources, e.g., many United States government agencies, especially subdivisions of the Department of Defense and the National Aeronautics and Space Administration. Producers of components and devices normally have information on failure rates and derating on their products, especially items which have been produced extensively. Such data are extremely limited on relatively new items and items produced in small quantities.

Derating—well-known and commonly practiced—is one of the most powerful tools available to the designer. It can contribute immensely to improving the reliability of a device and can play an important role in increasing the overall reliability of a complex system involving many components. There are limits, however, in terms of cost, weight, space, and similar considerations.

It should be emphasized that ratings (e.g., voltage, current, temperature power dissipation, etc.) are not clearly established limits above which immediate failure will occur and below which the device will operate indefinitely. Life of most devices increases in some continuous fashion as the stress level decreases below the rated level. Rating only represents a *judgment* of the stress level at which the life of the device is considered satisfactory or acceptable. In general, the greater the derating, the longer the life of the device. There is a minimum stress level, however, below which the increased system complexity will offset the gain in reliability.

Derating cannot be applied arbitrarily. There is danger in underrating as well as in overrating. One resistor of carbon composition, for example, has greatly increased life with operation at 50 percent of its rated wattage because it runs cooler with an accompanying decrease in resistance variation from chemical change. If derating is further increased, to perhaps 90 percent, there will be too little heat generated to drive out absorbed moisture. Chemical changes due to the absorbed moisture will occur more rapidly, resulting in shorter life. Some components must be operated close to rating. Electronic vacuum tubes, for example, are in this category. Reduced filament heater voltage shortens tube life, since the cathode operates at too low a temperature to keep its surface free of deposited impurities.

It appears wise to work closely with a component supplier to develop a coordinated program to determine desirable (and, it is hoped, optimum) operating parameters for a given application. Presumably, the supplier is in a better position to perform tests required to eliminate gaps in knowledge to meet complex system requirements.

If it is true, as assumed, that a "correct" device exists for every application, derating will gradually be deemphasized as the proper parts are developed. In the meantime, in a rather real sense, derating should be regarded as conservatism compensating to some extent for ignorance.

8-2 MAINTENANCE

All systems, except for nonrecoverable ones such as unmanned satellites, Mars Mariners, etc., which are used on a continuous or intermittent basis for any appreciable period of time, are subject to maintenance at some time or other. Properly maintained equipment can function at an effectively higher reliability than unmaintained equipment.

Maintenance actions are in two classes: (1) Off-schedule (corrective, "imperfect") maintenance required by system in-service failure or malfunction. In this case, the system is not serviced after each mission, but is allowed to proceed on the next mission, provided the system functions satisfactorily. Failed units are not replaced or repaired until the system fails. System operation is restored as soon as possible by replacing, repairing, or adjusting the devices which interrupted service. (2) Scheduled (preventive, "perfect") maintenance performed at regular intervals for the purpose of keeping the system in a condition consistent with built-in levels of reliability, performance, and safety. Scheduled maintenance, therefore, is performed to prevent component and system failure rates from exceeding design levels. For these reasons, it is known as preventive maintenance.

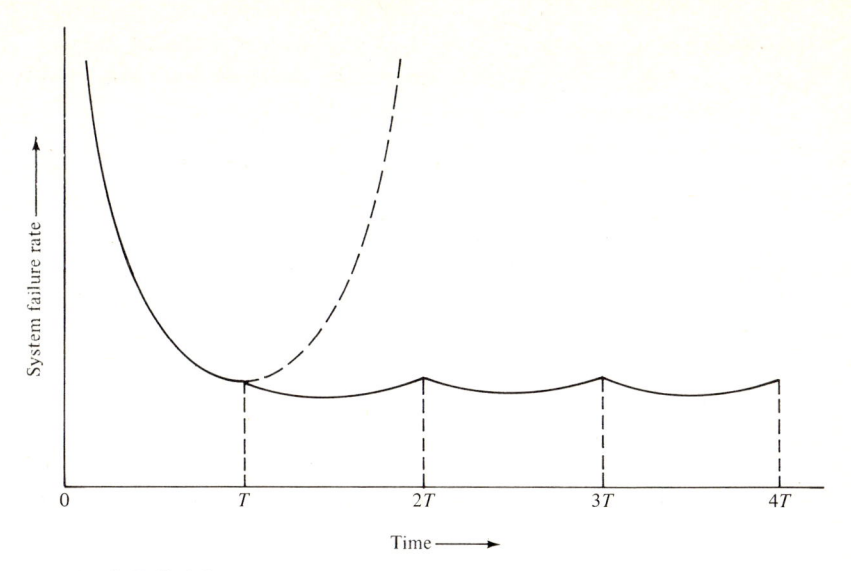

FIGURE 8-3
Effect of preventive maintenance at intervals of T hours on system failure rate, schematic.

Preventive maintenance achieves its purpose by inspections, servicing, and minor and major overhauls during which the actions fit into three classes:

1 Regular care of normally operating subsystems, devices, and components which require attention (lubrication, refueling, cleaning, adjustment, etc.).
2 Checking for, with replacement or repair of, failed redundant components.
3 Replacement or overhaul of components or devices which are nearing wearout.

The frequency of performing class 1 actions to prevent degradation of system reliability depends on the characteristics of the components. The frequency of performing class 3 actions depends on wearout characteristics and the number of components in a system. While these differ for various kinds of devices, an optimum replacement time table can be established in advance. The frequency of performing class 2 actions is a function of probabilities since it depends on failure rates of redundant components in a system and on the reliability required of the system.

An indication of the effect of preventive maintenance on system failure rate is shown in Fig. 8-3. It is obvious that periodic replacement of short-lived components greatly decreases the effective system-failure rate. This is typical of automotive and aircraft engines, many domestic appliances, machine tools, and other similar devices.

Periodic maintenance is performed every T h, starting at time zero. Each device and component is checked. Each one which has failed is replaced by a new, statistically identical component. For useful life, the system is restored to "as good as new" condition after each maintenance operation, since there has been no deterioration of components.

For periodic maintenance, the time period t can be written as

$$t = jT + \tau \qquad j = 1, 2, 3, \ldots \qquad 0 \leq \tau < T \qquad (8\text{-}9)$$

For a time period for which $j = 1$ and $\tau = 0$,

$$R_{PM}(t = T) = R(T)$$

If $j = 2$ and $\tau = 0$, the system has to operate for the first T h without failure of a redundant subsystem. After all failed elements are replaced, T h more of failure-free operation are necessary. Thus

$$R_{PM}(t = 2T) = [R(T)]^2$$

If $0 < \tau < T$, then τ h more of failure-free operation are necessary. Reliability is

$$R_{PM}(t = 2T + \tau) = [R(T)]^2 R(\tau)$$

In general,

$$R_{PM}(t = jT + \tau) = [R(T)]^j R(\tau) \qquad j = 1, 2, 3, \ldots \ 0 \leq \tau < T \qquad (8\text{-}10)$$

The MTTF [from Eq. (3-3)] for periodic maintenance is

$$\text{MTTF}_{PM} = \int_0^\infty R_{PM}(t)\, dt$$

The integral over the time range from zero to infinity can be expressed as the sum of integrals over intervals of T, i.e.,

$$\text{MTTF}_{PM} = \sum_{j=0}^\infty \int_{jT}^{(j+1)T} R_{PM}(t)\, dt$$

If $t = jT + \tau$, then $dt = d\tau$, and the limits of the integral are zero and T. As a result

$$\text{MTTF}_{PM} = \sum_{j=0}^\infty \int_0^T R_{PM}(t)\, dt = \sum_{j=0}^\infty \int_0^T [R(T)]^j R(\tau)\, d\tau$$

$$= \frac{\int_0^T R(\tau)\, d\tau}{1 - R(T)} \qquad (8\text{-}11)$$

EXAMPLE 8-1 Compare the MTBF, reliability for a 10-h mission, number of system failures (assume 1000 missions), number of system inspections, number of system checks, number of repair or replacement actions, total number of units required, and the costs involved for:

(a) A single unit with an MTBF of 100 h
(b) Three such units in parallel with off-schedule maintenance

(c) Three such units in parallel with "perfect" i.e., periodic maintenance after each 10-h mission
(d) A single unit having the MTBF of case (c)

MTBF

(a) Given as 100 h
(b) From Eq. (4-8)

$$\text{MTBF} = \frac{1}{\lambda} + \frac{1}{2\lambda} + \frac{1}{3\lambda} = (1 + \tfrac{1}{2} + \tfrac{1}{3})(100) = 183.3 \text{ h}$$

(c) $R = 1 - Q^3 = 1 - (1 - R_1)^3$
$= 1 - (1 - e^{-.1})^3$
$= 0.999138$
Expect 0.862 failure/1000 missions of 10 h each, i.e., 0.862 failure/10,000 h
MTBF = 10,000/0.862 = 11,600 h
(d) MTBF = 11,600 h

Reliability

(a) $R = e^{-\lambda t} = 0.90484$
(b) MTBF = 183.3 h
Expect one system failure for every 18.33 missions.

1000/18.33 = 54.56 system failures per 1000 missions $R = 0.94544$

(This is an "average." When all three units are good, $R = 0.999138$; when two are good, $R = 0.991$; when only one is good, $R = 0.90484$.)
(c) R (previously calculated) = 0.999138
(d) Same as (c).

Number of System Failures

(a) $N_f = NQ = 1000(1 - R) = 1000(1 - 0.90484) = 95.16$
96 system failures/1000 missions
(b) Calculated in determining reliability,
55 system failures/1000 missions
(c) Calculated in determining MTBF,
1 system failure/1000 missions
(d) Same as (c)

Number of System Inspections The implication is that the system will be inspected before each mission to ensure that it has not failed prior to the start of the mission. This is an inspection of the entire system. It is not an inspection or check of individual units. Therefore, there will be 1000 system inspections for all four cases.

Number of System Checks The implication is that if the system inspection indicates a failure of the system, all units will then be checked to locate the failed units.

(a) 96
(b) 55
(c) Only one system failure is expected in 1000 missions. This is attained by being certain that all three units are functioning before the next mission is started. We need the probability of one unit (or more) failing. From the binomial theorem, the probability of one or more failing is

$$P(1 \text{ or more}) = 3R^2Q + 3RQ^2 + Q^3 = 1 - R^3$$
$$= 1 - (e^{-\lambda t})^3 = 0.25918$$

Number of checks = 260/1000 missions
These 260 checks will show that one, two, or all three units have failed. (Using a failure indicator on the system does not tell how many units fail.) If a failure indicator is used on each unit, we can tell how many units fail.

$P(\text{one unit fails}) = 3R^2Q = 0.23374$
$\qquad = 234/1000 \text{ missions}$
$P(\text{two units fail}) = 3RQ^2 = 0.02458$
$\qquad = 25/1000 \text{ missions}$
$P(\text{three units fail}) = Q^3 = 0.00086$
$\qquad = 1/1000 \text{ missions}$
(Number of units failed = $234 + 2 \times 25 + 3 \times 1 = 287$)

(d) One system check

Number of Repair or Replacement Actions

(a) 96
(b) 55
(c) 287 units fail. 287 replacement actions. This is also the number of unit checks if indicators are used. If there are no indicators, all units must be checked before each mission; therefore, 3000 unit checks would be required.
(d) One

Number of Units Required (Assuming replacement rather than repair)

$\qquad$ Initial + Replacement
(a) $1 + 96 = 97$
(b) $3 + 165 = 168$
(c) $3 + 287 = 290$
(d) $1 + 1 = 2$

Costs Involved The cost factors which are involved in comparing the four cases are:

1 Cost of total number of units needed
2 Cost of making system inspections
3 Cost of making system checks
4 Cost of checking individual units

5 Cost of repair or replacement, including instrumentation and labor for diagnosis, repair or replacement, logistics, and function check after installation of "new" unit
6 Cost of servicing of units
7 Cost of downtime
8 Overhead cost
9 Miscellaneous administrative cost

Which of the four unit-maintenance combinations is best has to be judged in the context of the specific situation. This would not only include all the costs indicated above but the "penalty" which would accrue because of a failure during a mission. A cost analysis might well favor case (*d*), i.e., the single unit with an MTBF of 11,600 h. This would undoubtedly be a high-cost unit which might still need to be developed. The cost of such development might be more than met by the accompanying reduction in costs of items 2 through 9, above. Furthermore, the aspects of system weight and volume, in addition to accessibility for maintenance, might also favor such a single unit, given appropriate current state-of-the-art capabilities and sufficient development time.

Case	a	b	c	d
Initial no. of units	1	3 (imperfect maint.)	3 (perfect maint.)	1
MTBF, h	100	183.3	11,600	11,600
Reliability	0.90484	0.94544 ("average")	0.999138	0.999138
No. system failures/ 1000 missions	96	55	1	1
No. system inspections 1000 missions	1000	1000	1000	1000
No. system checks/ 1000 missions	96	55	260 (with system failure indicator) 287 (with unit-failure indicator)	1
No. unit checks/ 1000 missions	96	165	287 3000 (without indicator)	1
No. repair or replacement actions/1000 missions	96	165	287	1
Total no. units required/ 1000 missions	97	168	290	2

PROBLEMS

8-1 If the level of stress changes during a mission, then the failure rate also changes. At takeoff, for example, an aircraft engine has to generate a greater torque to get the higher engine thrust required. At cruising altitude and speed, torque requirements are reduced. Assume the stress profile of an aircraft flight is as shown.

90 INTRODUCTION TO RELIABILITY IN DESIGN

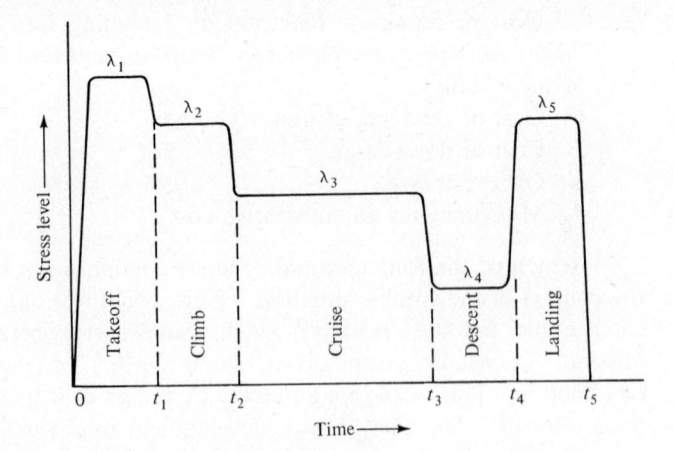

[Figure: Stress level vs. Time plot showing mission phases — Takeoff (λ_1) from 0 to t_1, Climb (λ_2) from t_1 to t_2, Cruise (λ_3) from t_2 to t_3, Descent (λ_4) from t_3 to t_4, and Landing (λ_5) from t_4 to t_5.]

(a) Find an expression for reliability of a single engine (useful life).

(b) Assume a four-engine aircraft. If all four engines are required for takeoff and climb, but only two out of four are required for completing the flight, determine the engine system reliability.

8-2 A device has a nominal failure rate of 10^{-6} failure/h. It is functioning at 40 percent of its rating so that K_{ap} is $\frac{1}{10}$. The environment is a relatively rigorous one with a K_{en} of 15. The component has a cyclic function which alternates from on to off and then back 200 times during an operating period of 1000 h. The total time for all switching is 1 h and the device has a failure rate of 10^{-5} failure/cycle. The device functions 65 percent of the time and is off the rest of the time (failure rate while deenergized is 10^{-7} failure/h). Calculate the reliability of the device for this operating period.

8-3 Problem 5-8 asked for a reliability of a standby system. There is an implication that each mission begins with all components in good condition with an assumed failure rate of zero of the battery while it is standing by. What happens to the MTBF and reliability of this standby system if the system is allowed to operate without maintenance until it has a random failure?

8-4 Four components (nonidentical) operate in parallel in a reliability sense. They have the following failure rates: (A) $\lambda_A = 0.003$ failure/h; (B) $\lambda_B = 0.005$ failure/h; (C) $\lambda_C = 0.002$ failure/h; (D) $\lambda_D = 0.006$ failure/h.

(a) Determine the reliability of the four-unit system for a 600-h operating period with perfect maintenance.

(b) How many of each unit will be required for 50 such operating periods?

(c) Determine the reliability of the four-unit system for a 600-h operating period with imperfect maintenance.

(d) How many of each unit will be required for 50 such operating periods?

(e) If a checkout is defined as determination of proper functioning of a new unit which has replaced a failed unit, how many checkouts will be required of each type of unit with perfect maintenance if there is a failure indicator on each unit?

(f) How many checkouts will be required of each type of unit with imperfect maintenance if there is a failure indicator on each unit?

(g) What assumption did you make as a definition for system success? What assumption did you make relative to parts (e) and (f)?

(h) Discuss the relative merits of perfect and imperfect maintenance.

8-5 An electronic device contains 2000 transistors as well as a large number of capacitors, diodes, and other electronic components. How many transistors would you expect to fail during a useful life operating period of 10,000 h if they are used at:

(a) A rated voltage at rated temperature of 80°C?

(b) 40 percent of rated voltage and 100°C?

(c) 120 percent of rated voltage and 40°C?

(d) 140 percent of rated voltage and 80°C?
(e) 40 percent of rated voltage and 90°C?
(f) 20 percent of rated voltage and 50°C?
(g) What rating level and temperature would you recommend for good practice?

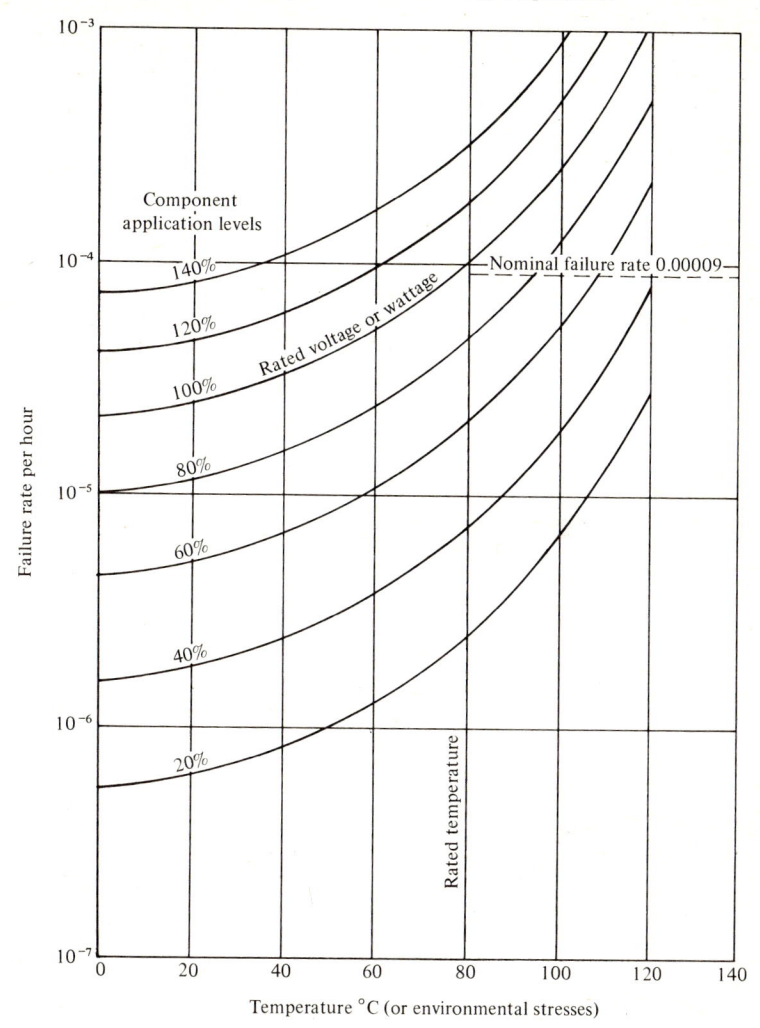

FIGURE P8-5
Derating curves for electrical and electronic components.

9
RELIABILITY TESTING

9-1 INTRODUCTION

Reliability tests measure or demonstrate the ability of equipment to operate satisfactorily for a prescribed period of time under specified operating conditions. The objective of a reliability test program is to gain information concerning failures, i.e., the tendency of systems to fail and the resulting effects of failure. Thus, in a sense, reliability tests are distinguished from most other types of tests, which are generally concerned with normal operation of equipment.

Few tests can be strictly categorized as "reliability" or "nonreliability." In a very wide sense, every test has an ultimate objective of improving product quality and any tendencies to failure will be noted whenever they are observed. Any test which relates to a malfunction or the resulting consequence can be considered a reliability test.

In our context, a reliability test has the general objective of obtaining information concerning failures, especially patterns of failure occurrence. As a result, the test may merely be a collection of data and analysis of activity relating to some other type of engineering test or with the operation of a system. The ideal reliability test makes use of all possible available test data.

9-1-1 Reliability Testing Objectives

Reliability tests can be considered in three classes: development and demonstration; qualification and acceptance; and operation.
Development and demonstration have the objectives of:

1 Determining if design improvement is needed to meet the reliability requirement.
2 Indicating if any design changes are needed.
3 Verifying improvements in design reliability.

Qualification and acceptance have the objectives of:

1 Determining if a part, assembly, subsystem, or other end item should be accepted or rejected (on an individual or lot basis).
2 Determining if a given design is truly qualified for its intended application.

Operation has the objectives of:

1 Verifying the reliability analyses previously performed.
2 Providing data which indicate necessary modifications or operational procedures and policies, especially as they influence reliability and maintainability.
3 Providing information which can be used in later activities.

Corollary objectives include:

1 Demonstrating satisfaction of contractual reliability requirements since a contract often includes an incentive clause whereby profit is readjusted (upward or downward) as a function of the relationship between demonstrated reliability and required reliability.
2 Determining a reasonable warranty period that will enhance sales while assuring reasonable profit.
3 Determining a corrective and preventive maintenance program required to keep the equipment in satisfactory operating condition.

9-1-2 Reliability Data from Other Tests

Reliability is only one of many characteristics of equipment. The other characteristics include such things as maintainability, interchangeability, structural integrity, and compatibility with other equipment in the system.
Data relative to these characteristics are commonly obtained during (and from) all tests. Failure-rate data, for example, may be obtained from laboratory-bench tests, evaluation and qualification tests, and factory-acceptance tests. Reliability tests are then designed to gather whatever data are still necessary to measure or demonstrate reliability with the desired degree of statistical confidence.

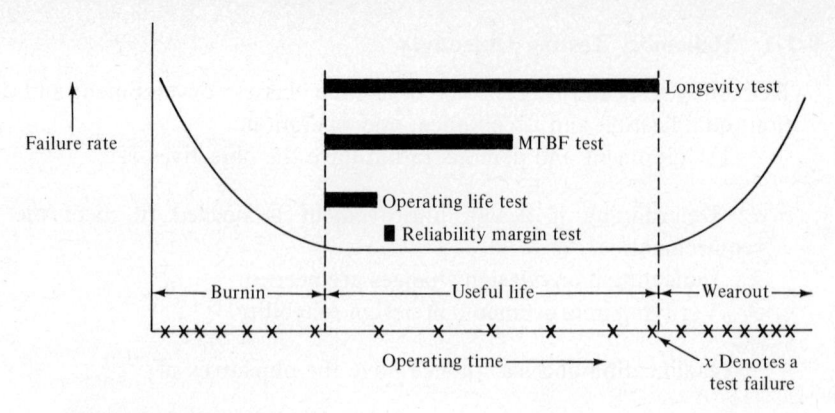

FIGURE 9-1
Types of reliability tests.

Equipment life naturally falls into three sequential phases as indicated earlier, namely:

1 Burn-in phase, during which there are high failure rates because of infant mortalities. This phase occurs during factory assembly and checkout prior to delivery to the customer.
2 Useful life phase, in which there is a statistically constant failure rate.
3 Wearout phase with a relatively high failure rate because of old-age mortalities.

Reliability tests are normally conducted during useful life since this is the period of interest to the customer (Fig. 9-1).

9-1-3 Types of Reliability Tests

1 Longevity tests measure or demonstrate the duration of the useful life phase, i.e., the constant failure rate phase (Fig. 9-1).
2 MTBF tests measure *m*ean *t*ime *b*etween *f*ailures. If equipment has a cyclic operation, mean cycles between failures (MCBF) may have more meaning (Fig. 9-1).
3 Operating life tests measure ability to perform without failure for a prescribed minimum period (Fig. 9-1).
4 Reliability margin tests measure the margin of safety between the extremes of operating environments and the limits of ability of the equipment to withstand these environments (Fig. 9-2).

9-2 TESTING PROGRAM

A reliability test program is a dynamic procedure which changes and evolves with the needs of a design and development effort. The first step is the preparation of a comprehensive plan detailing the specific requirements of each category

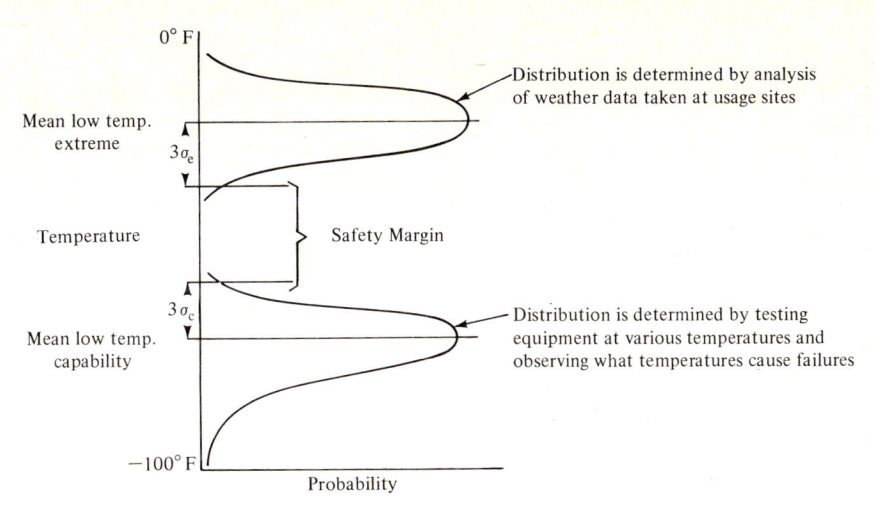

FIGURE 9-2
Reliability margin tests.

of testing and describing the testing responsibility of all organizations involved. This plan should have sufficient flexibility to allow modification as the program progresses.

A major input to preparation of the test plan is a review of available data pertinent to the specific equipment involved. Such data may satisfy some of the test objectives and also may pinpoint potential trouble areas.

9-2-1 Principal Elements of a Test

The major elements of a test are:

1. Time—When was the test conducted?
2. Place—Where was the test conducted?
3. Objective—Why was the test conducted?
4. Sample—What was tested?
5. Conditions—Under what conditions was the sample tested?
6. Performance criteria—What was considered to be acceptable performance?
7. Results—Was performance acceptable?
8. Recommendations—What corrective action, if any, is necessary?

9-2-2 Types of Tests

As equipment progresses through the design-development-production process, many tests are conducted. These can be considered in four classes: evaluation, qualification, service, and production. These test classes are time-phased as indicated in Fig. 9-3 and are separated from each other by design releases

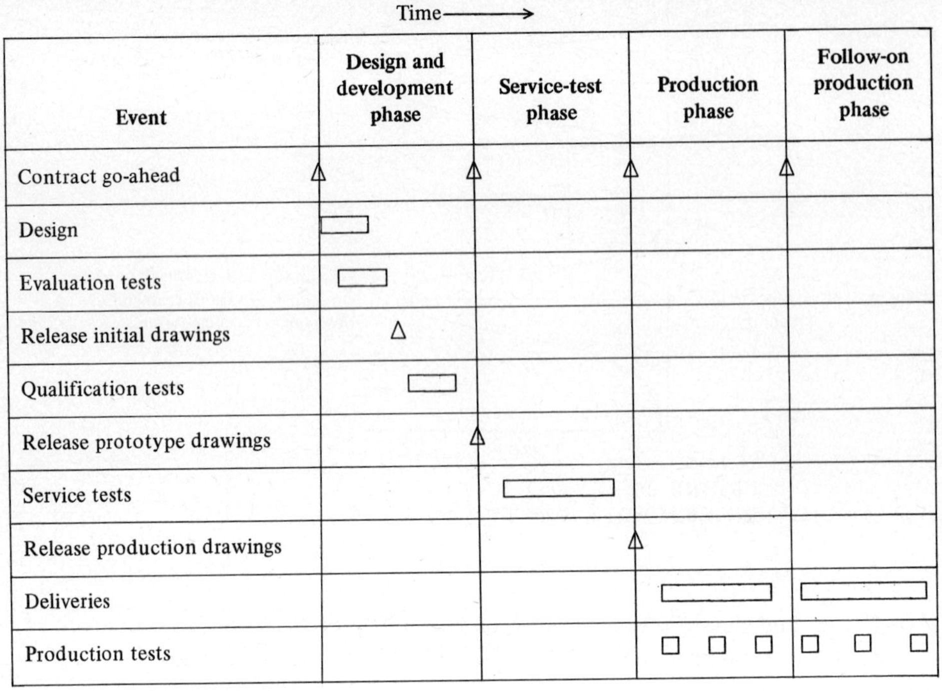

96   INTRODUCTION TO RELIABILITY IN DESIGN

Event	Design and development phase	Service-test phase	Production phase	Follow-on production phase
Contract go-ahead	△	△	△	△
Design	▭			
Evaluation tests	▭			
Release initial drawings	△			
Qualification tests	▭			
Release prototype drawings		△		
Service tests		▭		
Release production drawings			△	
Deliveries			▭	▭
Production tests			□ □ □	□ □ □

Time ⟶

FIGURE 9-3
Time-phased plan of testing program.

and/or contract changes. It should be recognized that Fig. 9-3 is somewhat idealized since development of some items in a system lags behind others, thereby causing overlaps. In addition, development programs may be partially expedited by taking some risk and deliberately overlapping some test phases.

The key elements of the tests change as the program advances, as shown in Fig. 9-4. Test objectives change from measurement of unpredictable performance to demonstration of satisfactory performance. Test responsibility shifts from engineering to factory to customer. Changes in equipment design are more rigorously controlled during the later phases of the program. Performance criteria are refined as the program advances, with some tolerances being tightened while others are loosened. Configuration of test samples changes:

Exploratory development models are breadboard and experimental models fabricated in the laboratory.

Advanced development models (prototypes) assume some appearances of production models and may be fabricated by engineering or factory personnel.

Service test models have the configuration of production models, are pilot production models, and are normally fabricated in the factory.

Aspect	Evaluation tests	Qualification tests	Service tests	Production tests	Reliability tests
Time	Design phase	Development phase	Development phase, pilot production phase, perhaps production phase	Production phase	All phases
Place	Engineering lab	Engineering lab	Customer's proving grounds	QC lab	Engineering lab, then QC lab
Objectives	Measure capabilities	Demonstrate capabilities	Demonstrate capabilities	Demonstrate capabilities	Measure or demonstrate reliability
Samples	Exploratory development models	Advanced development models	Service-test models	Production models	All models
Conditions	Bench; perhaps a few environments	Many environments	Actual field environments	Bench; perhaps a few environments	Bench; various environments
Performance criteria	Compliance with design objectives	Compliance with detail specification	Compliance with operational system requirement	Compliance with production specification	As applicable
Results	Design refinements; specification modifications	Finalize design; finalize specifications	Customer confidence; production go-ahead	Continued production	Incentive fees; profitable warranty; maintenance program

FIGURE 9-4
Key elements of tests.

Production models are factory-fabricated with "hard tooling," i.e., jigs, fixtures, and, other tooling that give high production rates.

Reliability tests are performed throughout the entire design-development-production process.

9-2-3 General Requirements of Test Procedures

Equipment reliability test procedures start with customer reliability requirements for the system using the equipment. In the case of the military, the requirements are usually specified in the contract or in military specifications invoked by the contract.

The time-phased program plan (Fig. 9-3) establishes the basic schedule of testing.

Sampling requirements are based on an equipment test sequence, as shown in Fig. 9-5 for a missile. Once parts, materials, and subcontracted items have been selected, tests of nonstandard (unqualified) items begin. The nonstandard items tested may be as many as 10 percent of the items required. They are tested separately, rather than as part of the system, to establish data which will qualify them for use in other applications. Assemblies and subsystems are then tested to assure success during the final (and most expensive) test of the entire system which completes the test program (until major changes occur).

9-2-4 Test Procedure Contents

A test procedure must be prepared for each nonstandard part, material, subcontracted unit, assembly, subsystem, and for the system itself. This procedure has a cookbook, step-by-step approach so that a given kind of test, performed in several places or by different people, will yield the same results. In addition, it provides a basis for estimating and scheduling the test program.

The test procedure prescribes test conditions, performance standards, number of samples, duration of test, and test sequence.

9-2-5 Test Conditions

Requirements, as written, are often intended for average equipment. These must be modified where necessary to suit individual equipment. (Incomplete resolution of this often leads to controversial negotiations.) Each system, subsystem, assembly, etc., must be considered individually to answer such questions as: What are the usage conditions to which it will be exposed? Will it be packaged or sheltered during each usage condition? Must it perform one or more functions during or after each exposure? What are the environments associated with each usage condition? Combining the answers to these questions will determine the environmental conditions that must be imposed during the test. A detailed outline of factors to consider is given in Appendix B.

RELIABILITY TESTING 99

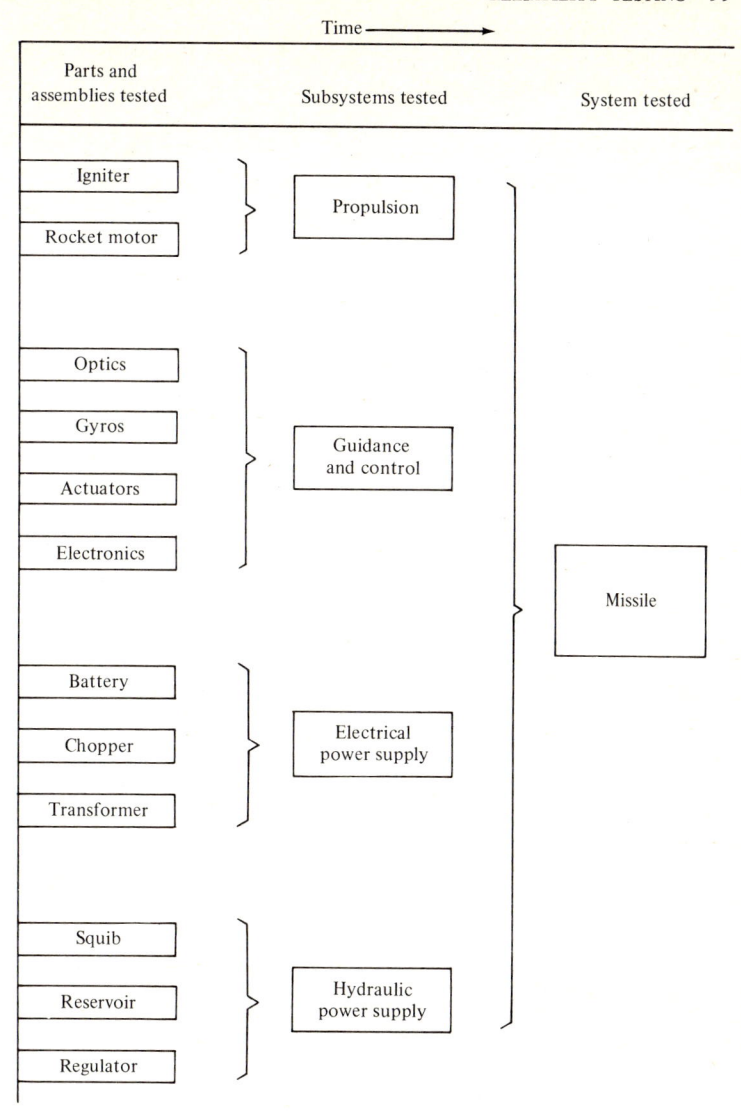

FIGURE 9-5
Equipment-test sequence for a missile.

Nonvariate or stationary life tests involve only one set of stress conditions for a specified time period. Many life tests for electronic parts and systems are in this class. These can be considered as rating verification procedures and appear regularly in procurement specifications.

Univariate life tests determine the performance of a device when one stress is varied at a time. A regular step increase can be applied to the items in test to determine separate behavior patterns as the major stresses, e.g., temperature, are increased. A variation of the procedure involves application of a controlled, continuously increasing stress. Such tests are very useful in determining stress levels at which various failure modes occur.

Multivariate life tests determine the behavior of items in test under the influence of various levels of the different stresses. Such tests utilize techniques of statistical design of experiments such as factorial and fractional factorial designs, response surface methodology, or evolutionary operation.

9-2-6 Performance Criteria

The equipment inputs and outputs, including nominal values and tolerances (ranges), must be defined. Any dummy loads required to simulate adjacent equipment must also be defined. *Failure criteria must be defined* including catastrophic, major, and minor failures. The allowable number of failures must be clearly determined and stated. It is also necessary to determine if performance is to be measured during or after environmental exposure, or both.

9-2-7 Sample Size and Test Duration

The sample size is determined by the desired confidence level but this is often modified, sometimes rather drastically, by the allowable cost and the time available to run the test program. An indication of the tradeoff involved is shown in Fig. 9-6.

Schemes for determining appropriate sample sizes for various conditions can be found in a variety of places, typical examples of which are given by Refs. 9-1 to 9-9. Despite these statistically correct sampling plans, the typical sample sizes for qualification tests are:

1 Individual parts—tens to hundreds
2 Units or subsystems—several up to ten or twenty
3 Systems—one to ten

Test durations usually range from hundreds to thousands of hours. The test configuration of each sample must be very carefully defined, taking into account all conceivable variables, or the test will be of limited (and possibly zero) value. Spare samples should be provided on the order of about 10 percent in an individual parts test—and perhaps one or two in a systems test. Ingenious and shrewd scheduling will help to achieve maximum usage of samples.

9-2-8 Apparent Tyranny of Numbers

Consider the situation in which the failure rate of an element is λ_i in useful life. The corresponding reliability [Eq. (3-5)] is

$$R = e^{-\lambda_i t}$$

RELIABILITY TESTING

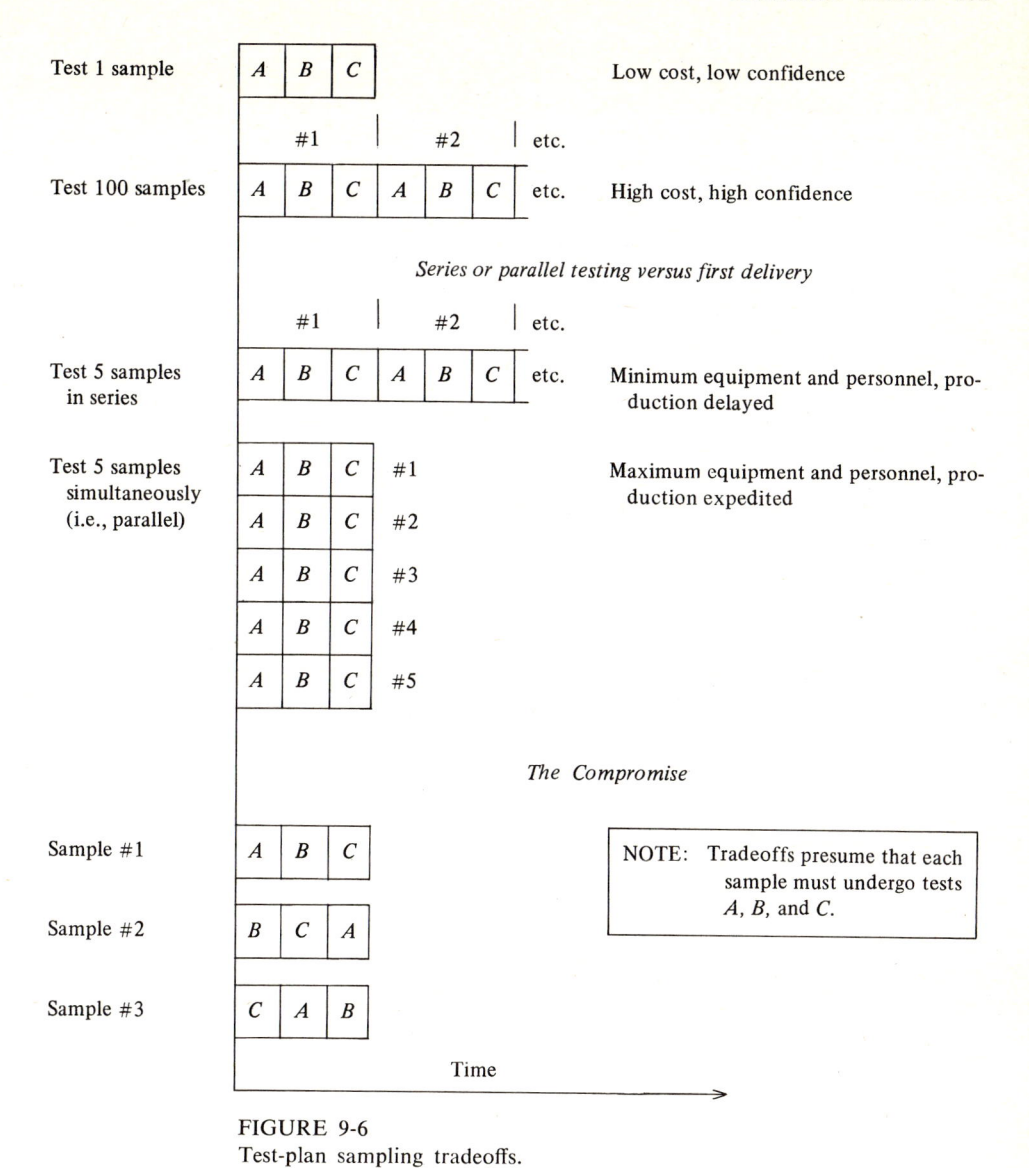

Test 1 sample	A	B	C				Low cost, low confidence

        #1   #2   etc.

Test 100 samples	A	B	C	A	B	C	etc. High cost, high confidence

Series or parallel testing versus first delivery

        #1   #2   etc.

Test 5 samples in series	A	B	C	A	B	C	etc. Minimum equipment and personnel, production delayed

Test 5 samples simultaneously (i.e., parallel)

A	B	C	#1
A	B	C	#2
A	B	C	#3
A	B	C	#4
A	B	C	#5

Maximum equipment and personnel, production expedited

The Compromise

Sample #1	A	B	C
Sample #2	B	C	A
Sample #3	C	A	B

NOTE: Tradeoffs presume that each sample must undergo tests A, B, and C.

Time →

FIGURE 9-6
Test-plan sampling tradeoffs.

If we have n such identical elements in series, then the reliability for the system is

$$R = e^{-n\lambda_i t}$$

The characteristic life θ [Eq. (3-3a)] is

$$\theta = \frac{1}{\lambda_i} = \frac{(nt)}{-\ln R}$$

If $n = 10$, $t = 10^3$ h, and $R = 0.99$, then

$$\theta = \frac{1}{\lambda_i} = \frac{10^4}{-\ln(0.99)} = 10^6$$

The characteristic life of each component should, thus, be 10^6 h. Since there are 8750 h/year, the characteristic life for this element should be about 115 years. To obtain a state of knowledge which assigns to λ_i a value less than 10^{-6} requires *more* than 115 years of *failure-free* testing (115 items for one year or 1150 items for six weeks) for *each* of the ten devices. Obviously, a great quantity of failure-free data must accumulate to offset failures which occur with a mean time less than 115 years.

If the number of parts is 10^3 and the reliability is required to be 0.999 (not at all unreasonable for a space mission), it is obvious that the resulting number of required tests becomes absurdly impossible.

This leads to a paradox known as the *tyranny of large numbers,* according to which it is impossible to accumulate sufficient data to guarantee that system-reliability requirements will be met. Yet we know that such systems are built and guaranteed!

The resolution of the paradox lies in the distinction between probability and frequency. Probability (and reliability) are states of knowledge, not states of things. In the above computations, the only data were prior knowledge and test data in the form of failure frequencies.

In any real situation, however, we have additional information which affects our knowledge. If a part fails, careful analysis and experimentation should reveal the cause of failure. A minor redesign may reduce the incidence of failure, for the given cause, to a negligible quantity. With the evidence relative to the redesign available, it is no longer necessary to accumulate vast times of testing to have a basis for determining a probability of failure from the given cause.

The point is that a careful analysis of possible failure modes, plus experimentation and exercise of engineering judgment which takes into account degradation, not just failures, can greatly reduce the total time required to establish that a probability of failure is less than the maximum permitted value.

9-3 PARAMETER ESTIMATION

Although one may wish to estimate various parameters in tests relating to reliability or parameters of distribution functions, a very common parameter which is desired is the characteristic life (or the corresponding failure rate). This is often done by conducting life tests in which a number of "identical" components or systems are operated in a carefully controlled environment until all (or a preassigned number) have failed. "Failure" must be very carefully and clearly defined in advance. The time for each component to fail is recorded. The test is terminated by the last failure.

9-3-1 Maximum Likelihood Estimator

There are a number of techniques for making good estimates. One of the better ones is the method of maximum likelihood estimation of distribution parameters developed by R. A. Fisher. The point estimates obtained from this technique are (statistically speaking) sufficient, asymptotically consistent (for large sample sizes, the parameter estimates converge to the true population parameters), asymptotically efficient (as sample size increases, the variance of the parameter estimates approaches a minimum variance), and approximately normally distributed.

Consider a sample of size n,

$$\{x_1, x_2, x_3, \ldots, x_n\}$$

drawn from a population described by the probability density function (pdf) $f(x;\theta)$ in which θ is the parameter of the population which we wish to estimate. We ask, "What is the probability of drawing this specific sample from all possible samples of size n?" That is, we wish to find the probability of a sample having values lying within the intervals $dx_1, dx_2, dx_3, \ldots, dx_n$.

The probability of one sample lying within dx_i at x_i is

$$f(x_i;\theta)\, dx_i$$

(from the definition of a pdf). The combined probability of the sample values lying within dx_1 at x_1, within dx_2 at x_2, and within dx_n at x_n is

$$[f(x_1;\theta)\, dx_1][f(x_2;\theta)\, dx_2] \cdots [f(x_n;\theta)\, dx_n]$$

provided that the observations are random and independent. This is equivalent to

$$\prod_{i=1}^{n} f(x_i;\theta)\, dx_i \qquad (9\text{-}1)$$

Eq. (9-1), with the differentials omitted, is known as the *likelihood* of the sample and can be considered to be a function of the parameter θ. To obtain a maximum, we take the partial derivative with respect to θ and set the result equal to zero. This will give the value of θ (call it $\hat{\theta}$) which provides a maximum likelihood. Since the pdf often involves exponential terms, it may be more convenient to work with the natural logarithm of the likelihood than with the likelihood itself. The logarithm of a function and the function itself will assume maximum values together. By definition

$$\mathsf{L}(\theta) = \ln \prod_{i=1}^{n} f(x_i;\theta) = \sum_{i=1}^{n} \ln f(x_i;\theta) \qquad (9\text{-}2)$$

Solving the equation

$$\frac{\partial \mathsf{L}(\theta)}{\partial \theta} = 0$$

for θ gives the desired maximum-likelihood estimate, $\hat{\theta}$.

$$\text{var } \hat{\theta} = \frac{\theta^2 n}{(n+1)^2(n+2)} \quad (9\text{-}2a)$$

The foregoing procedures are valid for use as a method for estimating parameters for any distribution. The procedures can be generalized to two or more parameters.

9-3-2 Application to Life Testing

In a life test, conducted as indicated above, the parameter of interest is the MTBF. It is quite common to assume the test is within the useful life period. (This assumption, while widely used, is not always valid.) For this situation, the pdf is

$$f(t;\theta) = (1/\theta)e^{-t/\theta} \quad t > 0; \theta > 0$$

where t is the variable time to failure and θ is the MTBF. The sample data from the test are

$$(t_1; t_2; \cdots ; t_n)$$

where

$$t_1 \leq t_2 \leq \cdots \leq t_n$$

The likelihood of this sample is

$$\left(\frac{1}{\theta}e^{-t_1/\theta}\right)\left(\frac{1}{\theta}e^{-t_2/\theta}\right) \cdots \left(\frac{1}{\theta}e^{-t_n/\theta}\right) = \frac{1}{\theta^n}\exp\left(-\frac{1}{\theta}\sum_{i=1}^n t_i\right)$$

from which

$$\mathsf{L}(\theta) = -n \ln \theta - \frac{1}{\theta}\sum_{i=1}^n t_i$$

$$\frac{\partial \mathsf{L}(\theta)}{\partial \theta} = -\frac{n}{\theta} + \frac{\sum_{i=1}^n t_i}{\theta^2} \quad (9\text{-}3)$$

equating to zero

$$0 = -\frac{n}{\theta} + \frac{\sum_{i=1}^n t_i}{\theta^2}$$

from which

$$\hat{\theta} = \frac{\sum_{i=1}^n t_i}{n} \quad (9\text{-}4)$$

giving the best estimate of the MTBF available from the test data.

9-4 ACCELERATED TESTING

It is obvious that a life test is generally expensive in both time and money. In many cases, failed units can not be repaired. Highly reliable units will require long operational (or testing) periods before useful failure data become available. Some means of speeding up testing is obviously desirable.

Accelerated testing can be accomplished by reducing the time required for testing by: (1) taking a large sample and testing only part to failure; (2) magnifying the load; or (3) sudden-death testing.

9-4-1 Partial Test to Failure

If a number of units are tested simultaneously, and $t_1, t_2, \ldots, t_i$ represent the recorded times of failure, it is obvious that

$$t_1 \leq t_2 \leq t_3 \leq \cdots \leq t_i \leq \cdots \leq t_n$$

In other words, failure will occur in an ordered sequence with the weakest failing first, etc. This ordering is unique to life testing.

If we are interested in estimating characteristic life, we may choose a sample of r items and wait for all r to fail, or we may choose a sample of n items $(n > r)$ and terminate the test after r of these have failed. The estimated characteristic life will have the same precision in either case but the latter test has a definite advantage over the former for the average waiting time for r out of n units to fail is less than the average waiting time for r out of r units to fail. This can also be seen intuitively, e.g., if there were one extremely long-lived item in the sample of r units and one in the sample of n units, we would not have to wait for it to fail in the latter case. Therefore, the choice of a test involving failure of r units out of n $(n > r)$ rather than choice of a test involving failure of r out of r units will, in general, permit determination of an estimate of characteristic life in a relatively shorter time. The average saving of time is indicated in Table 9-1. In many cases the $n - r$ units that have not failed will still be serviceable. If the failures follow a Poisson distribution, the $n - r$ survivors will be as good as new. Even if the survivors are considered to have deteriorated enough to render them unfit for further service, the appreciable saving in time may be well worth the cost of the additional units.

For this truncated case, the likelihood function is the joint probability of n-independent items failing and m (i.e., $n - r$) independent items *not* failing. The probability of an item *not* failing by time t_j is the complement of the probability that it *will* fail by time t_j. This probability is equal to $1 - F(t_j; \theta) = R(t_j; \theta)$. The joint probability can be expressed as:

$$J(\theta) = \prod_{i=1}^{n} f(t_i; \theta) \left[\prod_{j=1}^{m} R(t_j; \theta) \right] \qquad (9\text{-}5)$$

Table 9-1 RATIO OF EXPECTED WAITING TIMES TO OBSERVE THE rth FAILURE IN SAMPLES OF SIZE n AND r, RESPECTIVELY

r \ n	1	2	3	4	5	10	15	20
1	1	0.50	0.33	0.25	0.20	0.10	0.067	0.050
2	—	1	0.56	0.39	0.30	0.14	0.092	0.068
3	—	—	1	0.59	0.43	0.18	0.12	0.087
4	—	—	—	1	0.62	0.23	0.14	0.104
5	—	—	—	—	1	0.28	0.18	0.125
10	—	—	—	—	—	1	0.35	0.23

To interpret Table 9-1, we define

$E(t_{r,n})$ = average waiting time to observe first r failures from a sample size of n ($n > r$)

$E(t_{r,r})$ = average waiting time to observe all r failures from a sample of size r

The entries in the body of the table are values of the ratio $E(t_{r,n})/E(t_{r,r})$ and thus provide a quantitative measure of the time saved for specific choices of r and n.

For example, if we have to operate 10 units for 200 h to induce all 10 failures, we shall have to wait, on the average, only $(0.23)(200) = 46$ h to observe the first 10 failures in a sample of 20.

The likelihood function then becomes (by taking the logarithm)

$$\mathsf{L}(\theta) = \sum_{i=1}^{n} \ln f(t_i;\theta) + \sum_{j=1}^{m} \ln R(t_j;\theta) \qquad (9\text{-}6)$$

For the case of useful life and a set of sample data

$$\{t_1, t_2, \ldots, t_i, \ldots, t_n : t_1, t_2, \ldots, t_j, \ldots, t_m\}$$

where t_i = ith ordered failure time
t_j = jth ordered nonfailure time
n = number of failures
m = number of nonfailures

$$J(\theta) = \prod_{i=1}^{n} \lambda e^{-\lambda t_i} \prod_{j=1}^{m} e^{-\lambda t_j} \qquad (9\text{-}7)$$

but $\lambda = 1/\theta$; and taking logarithms

$$\mathsf{L}(\theta) = -n \ln \theta - \frac{1}{\theta} \sum_{i=1}^{n} t_i - \frac{1}{\theta} \sum_{j=1}^{m} t_j \qquad (9\text{-}8)$$

Differentiating with respect to θ

$$\frac{\partial \mathsf{L}(\theta)}{\partial \theta} = -\frac{n}{\theta} + \frac{1}{\theta^2} \sum_{i=1}^{n} t_i + \frac{1}{\theta^2} \sum_{j=1}^{m} t_j$$

equating to zero

$$\hat{\theta} = \frac{\sum_{i=1}^{n} t_i + \sum_{j=1}^{m} t_j}{n} \qquad (9\text{-}9)$$

It might be noted that when there are no survivors Eq. (9-9) simplifies to Eq. (9-4), which is the arithmetic mean of the failure times.

The foregoing calculations imply a nonreplacement test, i.e., no replacement of failed parts. Actually this can be done in two ways. If the test is terminated upon failure of a specified item (e.g., the rth item out of a set of n), then Eq. (9-9) becomes

$$\hat{\theta} = \frac{\sum_{i=1}^{r} t_i + (n-r)t_r}{r} \qquad (9\text{-}10)$$

If the test is terminated after a specified number of hours, t^*, when some have failed and others have not, i.e., $t^* > t_r$ then Eq. (9-9) becomes

$$\hat{\theta} = \frac{\sum_{i=1}^{r} t_i + (n-r)t^*}{r} \qquad (9\text{-}11)$$

If a replacement test is used, i.e., there is replacement of a failed part by a good one upon failure, then Eq. (9-9) becomes

$$\hat{\theta} = \frac{nt_r}{r} \qquad (9\text{-}12)$$

In the case of censored items (withdrawal or loss of items which have not failed), if failures are replaced and censored items are not replaced

$$\hat{\theta} = \frac{\sum_{j=1}^{c} t_j + (n-c)t^*}{r} \qquad (9\text{-}13)$$

where t_j = time of censorship and c = number of censored items.

If neither failures nor censored items are replaced,

$$\hat{\theta} = \frac{\sum_{i=1}^{r} t_i + \sum_{j=1}^{c} t_j + (n-r-c)t^*}{r} \qquad (9\text{-}14)$$

In terms of the Weibull distribution, the median time required to fail r out of n units as a fraction of the time required to fail n out of n (assuming all n can be run simultaneously and there are no replacements) is

$$F_T = \frac{T_r}{T_n} = \left\{ \frac{\log\left[1 - \lambda_n(r)\right]}{\log\left[\dfrac{0.69315}{r}\right]} \right\}^{1/b} \qquad (9\text{-}15)$$

where $\lambda_n(r)$ is the median rank of the rth value in n, given by

$$\lambda_n(r) = \frac{r - 0.30685 - 0.3863(r-1)/(n-1)}{n} \qquad n > 20$$

$$\lambda_n(r) = 1 - 2^{-1/n} + \left(\frac{r-1}{n-1}\right)\{2^{[1-(1/n)]} - 1\} \qquad n \le 20$$

T_r = median time to fail r out of n
T_n = median time to fail n out of n

If replacements are made as failures occur until there are r failures, then

$$F_T = \frac{T_r}{T_n} = \left(\frac{r}{n}\right)^{1/b} \qquad (9\text{-}16)$$

where b = Weibull slope
n = number of specimens running simultaneously
r = number of specimens to be failed

9-4-2 Magnified Loading

Use of magnified load does reduce testing time and possibly the number of items required for test. A major problem is that of correlation. For example, if we wish to know the performance of an engine in normal use for 5000 h, we can get much the same performance in 2830 h at full throttle, or in 100 h at 23 percent overload. This correlation is possible, since much information exists. In many situations, however, establishing such correlation is difficult, since we must first know what normal means and then we must have enough overload data to correlate with normal.

A good example of the effect of magnified loading has been developed by Conover, Jaeckel, and Kippola (9-10). In standard fatigue testing, a series of constant load amplitude tests are performed and the corresponding times to failure are recorded. This permits plotting standard *S-N* curves. It is recognized that few components in field service are subjected to constant amplitude loading. Programmed fatigue tests, developed from field-loading data, led to the development of mixed-cycle fatigue tests. If such a programmed fatigue test is intensified, i.e., all steps of the load program are increased by the same factor, there will be a shorter life to failure. If a series of tests with different scale factors are run, there will be different lives to failure. These results plot as a straight line on log-log coordinates. Experience has shown that the slope of this line changes very little for unnotched specimens, for notched specimens, and for automotive components. The slope for steels used in automotive components has a value k between 6.5 and 7.0. It should be noted that, while generalization is possible on the slope of this curve, there is no generalization possible on the form of the complete curves themselves. Having obtained failure life (N_1 cycles) in the laboratory at a given intensity factor (σ_1/σ_2) the

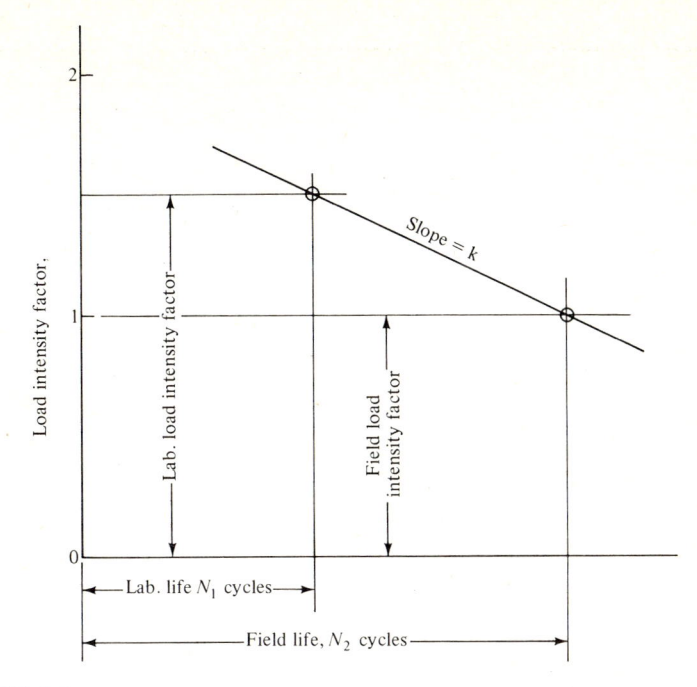

FIGURE 9-7
Schematic diagram of effect of intensified loading in predicting fatigue life from mixed-cycle *S-N* curves.

field life (N_2 cycles) at a scale factor of one (1) can be obtained from the relationship

$$N_2 = N_1 \left(\frac{\sigma_1}{\sigma_2}\right)^k \quad (9\text{-}17)$$

This is shown schematically in Fig. 9-7.

9-4-3 Sudden Death

Sudden-death testing applies especially to tests in which there are a large number of relatively inexpensive units. The method is not suitable for tests of a few units which are relatively expensive. Sequential testing is more useful in that case.

Consider a collection of fifty specimens which are available for testing. These fifty are randomly divided into ten sets of five items each. Each set of five is run simultaneously until one item in the set fails. After failure, the rest of the set is removed from test. This continues until all ten sets have been tested.

Ten numbers are obtained. Each one represents the least value in a ran-

110 INTRODUCTION TO RELIABILITY IN DESIGN

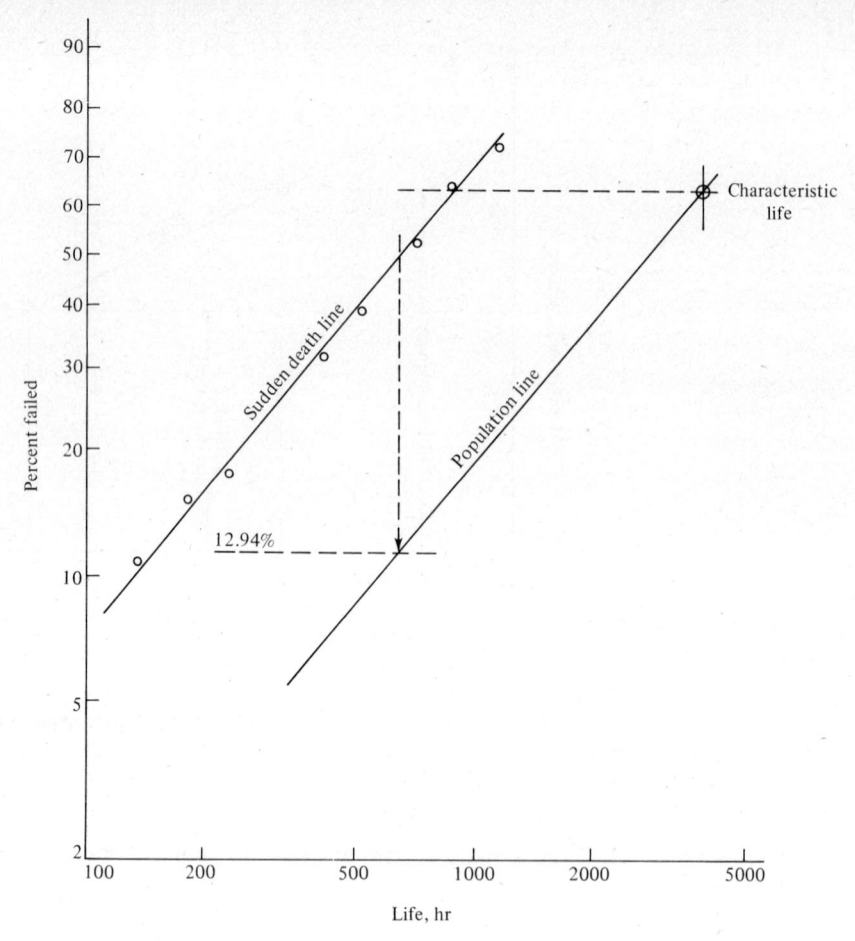

FIGURE 9-8
Schematic diagram of sudden-death parameter estimation.

dom collection of five. These ten numbers represent separate estimates of the life (see Appendix C) of the weakest 12.94 percent of the population.

The ten numbers are plotted on Weibull paper (Fig. 9-8) using the rank column for a sample of ten. A straight line is obtained which describes the distribution of the 12.94 percent life instead of the distribution of the life of entire population. As a best estimate of the 12.94 percent life point, the median (not mean) 50 percent life is taken. This is located at the 50 percent level of the straight line obtained.

This estimate of the 12.94 percent life is just as reliable as one obtained from testing all 50 specimens. In other words, sudden-death testing extracts all possible information about any one level of failure (for this example, the 12.94 percent level).

Since we now know the median life for 12.94 percent of the population, we can find this point on the Weibull plot. A straight line, drawn through this point, parallel to the sudden-death line, is used as a population line. This procedure is based on the assumption that the slopes of the population line and the sudden-death line are the same. This assumption may or may not be valid. The characteristic life is that value associated with the 63.2 percent point on the population line.

The fraction of time required to test r sets of specimens each (serial successive testing of one set at a time), relative to the time required to test all to failure ($n = rs$), is

$$F_T = \frac{T_r}{T_n} = r\phi \left[\frac{(1/s) \log 2}{\log [1/\lambda_n(1)]} \right]^{1/b} \qquad (9\text{-}18)$$

where $\lambda_n(1)$ is the mean rank of one failure out of n specimens and

$$\phi = \frac{(r-1)!\Gamma[1 + (1/b)](r - 1 + \ln 2)^{1/b}}{(\ln 2)^{1/b}\Gamma[r + (1/b)]}$$

If all sudden-death sets can be tested simultaneously

$$F_T = \frac{T_r}{T_n} = \left[\frac{(1/s) \log (1/[\lambda_r(1)])}{\log (1/[\lambda_n(1)])} \right]^{1/b} \qquad (9\text{-}19)$$

9-5 SEQUENTIAL TESTING

Sequential sampling and testing is an extension of multiple sampling in that samples are tested in sequence with the test result reviewed after each test is completed. Then the decision is made to accept, reject, or continue sampling. Under these conditions, the sample size is established only after an "accept" or "reject" decision has been made.

Assume we are interested in the reliability of a certain item of equipment. We desire a certain reliability R_d so that the item will perform satisfactorily. We are willing, however, to accept a somewhat lower reliability R_m, provided there is a high probability that lots with less than R_m will be rejected. We need four parameters:

R_d = desired or specified reliability
R_m = minimum acceptable reliability
α = producer's risk—probability of rejecting a lot whose reliability is equal to or greater than R_d
β = consumer's risk—probability of accepting a lot whose reliability is equal to or less than R_m

As sampling and testing progress, the number of failed items is plotted against the number of successful items (Fig. 9-9). Testing is continued until the plotted step function crosses one of the two decision lines. The decision lines

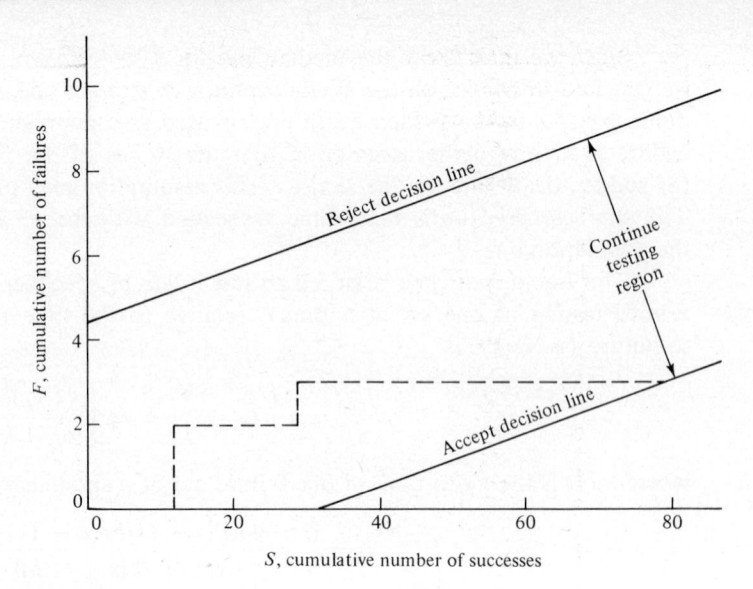

FIGURE 9-9
Schematic diagram of sequential testing procedure.

are obtained as follows:

$$\text{Accept:} \quad F \ln \frac{1 - R_m}{1 - R_d} + S \ln \frac{R_m}{R_d} = \ln \frac{1 - \beta}{\alpha}$$

$$\text{Reject:} \quad F \ln \frac{1 - R_m}{1 - R_d} + S \ln \frac{R_m}{R_d} = \ln \frac{\beta}{1 - \alpha} \tag{9-20}$$

where F represents cumulative number of failures and S cumulative number of successes.

It might be thought that a sequential testing plan could lead, on occasion, to an interminable test. It can be shown, however, that the test will eventually terminate. In fact, sequential testing will generally require testing of fewer items, on the average, than single or multiple sampling.

In a sequential test, the exact sample size is never known until the test has terminated. It is possible, however, to compute an average sampling number (ASN) once the choices for R_d, R_m, α, and β have been made. This number is

$$\text{ASN} = \frac{(1 - \alpha) \ln [\beta/(1 - \alpha)] + \alpha \ln [(1 - \beta)/\alpha]}{(1 - R_d) \ln [(1 - R_m)/(1 - R_d)] + R_d \ln (R_m/R_d)} \tag{9-21}$$

9-5-1 Application with Weibull Distribution

The use of Weibull distribution in sequential testing really involves a comparison test in which three decisions are made:

D_I: The new item is from a population with $\mu > \mu_\text{I}$ with confidence $(1 - \alpha)$ percent or greater

D_{II}: The new item is from a population with $\mu < \mu_{II}$ with confidence $(1 - \beta)$ percent or greater

D_{III}: Insufficient information, continue testing where
I—distribution of a standard material
II—distribution of minimum acceptable material (sample must not lie below this)

α = producer's risk
β = consumer's risk

1 To accept D_I, the following inequality must be satisfied.

$$\sum_{i=1}^{r} t_i^b > \frac{\theta_{II}^b}{\gamma^b - 1}\left[br \ln \gamma + \ln\left(\frac{1-\beta}{\alpha}\right)\right]$$

2 To accept D_{II}, the following inequality must be satisfied.

$$\sum_{i=1}^{r} t_i^b < \frac{\theta_{II}^b}{\gamma^b - 1}\left[br \ln \gamma - \ln\left(\frac{1-\alpha}{\beta}\right)\right]$$

3 Accept D_{III} if neither D_I nor D_{II} are accepted.

where r = number of failures

$\gamma = \dfrac{\theta_{II}}{\theta_I}$

b = Weibull slope

PROBLEMS

9-1 A set of 425 forty watt, 110 V, internally frosted incandescent lamps is tested for life. The frequency table for length of life is given.

Class interval (h)	Frequency
201–300	1
301–400	—
401–500	—
501–600	4
601–700	12
701–800	23
801–900	46
901–1000	90
1001–1100	86
1101–1200	79
1201–1300	46
1301–1400	25
1401–1500	10
1501–1600	2
1601–1700	1
	425

(a) Determine the mean, median and mode.
(b) What is the standard deviation of the sample?
A bank of 2800 such bulbs is constructed for a continuously operating heating-drying-baking oven. This is put into operation on a Monday morning.
(c) After four full weeks of operation, a maintenance man plans to replace all burned-out lamps. How many lamps should he draw from stock for replacements?
(d) If the oven cannot properly fulfill the requirements if more than 10 percent of the bulbs are burned out, what is the maximum time that should be allowed between replacements?
(e) How many bulbs would you expect to replace each year?

9-2 Twenty small generators were put under test for a period of 1500 h. One generator failed at 400 h and was replaced by a new one. A second failed at 500 h and was also replaced. A third and fourth failed at 550 and 600 h, respectively, and were removed from testing, but were not replaced. A fifth malfunctioned at 700 h, was immediately repaired, and was put back into test. A sixth malfunctioned at 800 h, but was kept in test. Later analysis showed this failure was due to governor malfunction. Estimate the failure rate of the generators. What assumptions did you make?

9-3 An engineer is testing a lot (100) of transmissions which are in useful life range. At the end of 1000 h, the total number of failures is six. Determine the following:
(a) The expected characteristic life θ
(b) The expected R_{50} life
(c) The expected R_{90} life†

9-4 Five gears were tested to failure, and the following lives were recorded: 2.0×10^5, 0.50×10^5, 1.7×10^5, 3.2×10^5, 0.90×10^5.
(a) Determine the Weibull slope and 90 percent confidence limits on this slope.
(b) Determine Weibull mean and 90 percent confidence limits on this mean.
(c) Determine the R_{90} life and 90 percent confidence limits on this life.
(d) Determine the characteristic life.

9-5 A shaft is manufactured with a large number of stress raisers. There is considerable doubt as to whether it can withstand the required load for 6×10^5 cycles as required. Ten shafts were tested to failure. The shafts failed at the following number of cycles: 3.5×10^5, 6.5×10^5, 8.0×10^5, 9.2×10^5, 1.3×10^6, 1.45×10^6, 1.68×10^6, 1.80×10^6, 1.95×10^6, 2.40×10^6. Using a Weibull plot, determine the following with 90 percent confidence:
(a) What can you say about the population that this sample represents? (Show graphically.)
(b) Give the range of percent of shafts that will fail before 6×10^5 cycles.

9-6 In order to evaluate the fatigue life of transmission gears, seven gears were picked at random out of a production run and fatigue tested at 55,000 psi, with the following results:

Life, cycles $\times 10^6$
1.7
1.0
1.5
1.2
1.8
2.0
2.3

(a) How many gears do you expect to have a life lower than the lowest life in the above sample?
(b) What is the R_{90} life?

9-7 Ten (10) gears were fatigue tested to failure and the life cycles at failure were recorded as follows:

† R_{90} life is life expected for 90 percent reliability.

RELIABILITY TESTING 115

Life to failure, cycles	
1.5×10^6	2.5×10^7
3.9×10^6	2.5×10^6
1.1×10^7	5.2×10^6
5.9×10^5	1.8×10^7
9.0×10^6	7.4×10^6

Using Weibull distribution determine the following:
(a) The mean life of the sample and the 90 percent confidence limit for the true mean.
(b) The R_{90} life of the sample and the 90 percent confidence limit for the true R_{90} life.
(c) The Weibull slope and the 90 percent confidence limit for the true slope.

9-8 In the course of a reliability analysis of a submarine fire-control system, it was decided to life-test a group of fuses: 500 fuses were put on test at 85 percent of rated current. After about 1800 h, 28 fuses had blown and the test was stopped.

A blown fuse was counted as a failure since fuses are expected to last indefinitely at 100 percent of rated current. If the fuses had been subjected to perhaps 120 percent of rated current, then those that had not blown by some specified time would have been counted as failures. The times (to the nearest hour) at which failures occurred are:

115	890	1405	1487	1590	1680
120	1160	1415	1505	1599	1700
205	1188	1437	1550	1650	1710
840	1300	1449	1575	1670	
848	1380	1489	1585	1675	

Calculate the characteristic life (MTTF, MTBF). Remember that the usual relationship assumes each failure is a random event following a Poisson distribution.

9-9 Seventy-two bearings are available for life testing. They are divided into six groups of twelve bearings each and tested by the sudden-death technique. The resultant data are given below:

Group 1: Bearing # 3 fails at 110 h
Group 2: Bearing # 4 fails at 75 h
Group 3: Bearing #11 fails at 165 h
Group 4: Bearing # 5 fails at 310 h
Group 5: Bearing # 7 fails at 210 h
Group 6: Bearing #12 fails at 270 h

What is the R_{90} life of the bearing population? Estimate the time required for 75 percent of the population to fail.

9-10 An engine manufacturer wishes to test a new style of exhaust valve on a V-8 engine. Five engines are put in test. The first valve fails in each engine as follows:

Engine no.	Time, h
1	170
2	75
3	220
4	125
5	320

Estimate the median life of these valves.

9-11 A maker of bearings has a unit which has a Weibull life distribution with a slope of 2.0 and a characteristic life of 600 h under standard testing conditions. A small redesign coupled with a change in material leads the maker to believe that he has doubled the characteristic life. He then proceeds to test. He wishes to know (with 95 percent confidence) which characteristic life is the more reasonable estimate for the new bearing. Given:

Bearing no.	Time to failure, h
1	700
2	1300
3	1100
4	900
5	1000

9-12 Determine by sequential analysis anything you can concerning the characteristic life of a component: 300 h characteristic life is highly satisfactory; 150 h characteristic life is unsatisfactory. Assume a Weibull slope of 1.5 (known from previous experience). A confidence of 95 percent is desired for your answer. Use the test data indicated below as necessary.

r	x
1	150
2	120
3	160
4	175
5	155
6	150
7	160
8	160
9	155

10
BROAD GUIDELINES FOR DESIGN

10-1 INTRODUCTION

In designing for reliability, it must be recognized that equipment must withstand *actual,* not *expected,* service conditions for the specified mission time. These actual service conditions are often not really known until prototypes are built and field-tested. In any case, the design problem involves prediction and probability with some sort of a scatter pattern of failure. Three major considerations are loading, environment, and material behavior. Time is the major independent variable. Loading commonly varies with time, while environment and material behavior are influenced by it.

10-2 APPROACHES TO THE PROBLEM

The designer, having the problem of meeting specified reliability requirements, should constantly search for alternative ways of achieving the requirements. Obviously, if performance requirements cannot be met, there is no point in going further. The requirements can generally be met, but the designer should never be satisfied with only one idea on how this may be accomplished. This one idea may give the least reliable equipment for the task or it may be one that makes it relatively difficult to achieve desired reliability. It is an old axiom that the simplest is often the best.

A given problem can be approached on either a "fail-safe" or a "one-hoss shay" basis. The fail-safe basis has a deliberately incorporated weak spot. It depends on the weakness to act in the same manner as a fuse in an electrical circuit. When the "fuse" fails, it can be readily replaced. The one-hoss shay basis makes all components equally strong. If weight or cost must be minimized, this approach—with all components having equal resistance to loading—may be appropriate. In practice, it is difficult to achieve this equality, since components fail in differing ways and at differing rates. The two bases indicated above, do *not* eliminate the necessity of considering alternatives within each approach.

The "worst case" approach is sometimes used. This effectively considers the worst possible set of circumstances or combination of variables and bases the design on this combination. While this is one valid procedure, it often results in overdesign. It is usually preferable to accept a set of service conditions with the accompanying probabilities and proceed.

10-3 LOADING

Loading can be externally applied. It can also result internally from component inertia or it can be induced by thermal effects. Loads can be measured or calculated. It is highly desirable that loading be determined by alternative methods. If this is not done, it is helpful to know the method used in order to have some estimate of the accuracy.

Maximum loading alone is not sufficient. At least as important is a knowledge of the rate and frequency of loading. A load spectrum gives relative frequency of amplitude but no indication of rate or frequency of application. A continuous load-time curve will give this information. It is rarely possible to know the number or extent of loads with high precision. It is generally possible, however, to obtain probabilities of occurrence.

Fatigue is an ever-present possibility. It should never be ignored in designing equipment.

10-4 ENVIRONMENT

Chapter 8 gave some consideration to the effects of environment with the discussion implying the effects of operating environment. It must be recognized that there are also three preoperational environments:

1. Before-assembly individual environments.
2. Internal environment generated by the equipment.
3. Shipping and storage environment.

These three preoperational environments can impose stresses on the equipment which must be considered in the design. In fact, in some cases, these environments may impose greater stresses than normal operation. Domestic appliances during shipping, for example, may be subject to shock, humidity, and temperature changes greatly exceeding those encountered in normal household use.

Operational environments include: shock; vibration; humidity; corrosive gases or liquids; ambient temperature; temperature gradients; surface temperature; surface finish; component geometry; available coolants, their temperature, flow rate, and other characteristics; sand and dust; fungi; radiation; acoustical noise; electrical noise; electrical fields; magnetic fields. Magnitude, duration, rate of change, and interdependence effects must be known for each. Time characteristics of environment affect design (and reliability), as well as average or maximum values.

10-5 MATERIAL BEHAVIOR

Materials properties are very sensitive to many environmental variables such as time, temperature, humidity, ambient chemicals, etc. Many of these interact. For example, consider a metal bar at room temperature with a relatively small static tensile loading which produces only elastic deformation. If the temperature is increased, the bar may elongate in plastic flow (known as creep). If a corrosive atmosphere (even a rather mild one) is added, the bar may creep even more rapidly.

Consideration of the ramifications of material behavior is far beyond the scope of the intended discussion. It must be recognized that this aspect must be very carefully accounted for and solved in the design. To overlook the materials aspect could well be catastrophic.

10-6 THREE BROAD STEPS

Good design for reliability is predominantly experience, therefore:

1 Carefully study similar past cases and their performances to learn as much as possible.
2 Give careful attention to the entire design but focus especially on variations and innovations in the new design.
3 Be conservative in initial specifications. Ratings and specifications can be modified in the appropriate directions and amounts as experience is accumulated.

10-7 ENSURING RELIABILITY

The concept of reliability places added responsibility on the designer. The *design* may meet all the requirements but the *product* may not. Therefore, the designer must be concerned with proper manufacture as well as design. He must:

1 Be specific on all tolerances and tests. These must eliminate defective units and devices as early as possible.

2 Be specific on materials characteristics. Vendor's tradenames are not sufficient. They must be supplemented by tolerances on important characteristics as the vendor may change the process or a characteristic without changing the tradename or designation.

3 Be specific on manufacturing processes with ample detail. In some cases this may be the only way to ensure desired characteristics of a device.

4 Be specific on sources of supply. Do not depend on the purchasing department to detect differences described above. If "identical" items are obtained from a different vendor, the items supplied by the new vendor must be tested and qualified.

5 Be specific on installation, maintenance, and use procedures as all three have major influence on reliability. A reliability group, if one exists, can (and should) advise, suggest, check, and test. The ultimate responsibility for inherent reliability, however, belongs to the designer.

6 Simplify equipment and components. Reduce the required number of parts and adjustable items to a minimum. Restrict adjustments to attaining the necessary range. Simplify inspection and checking procedures. Remember that the simpler the equipment necessary to meet given requirements, the better the design.

7 Select inherently reliable components to the greatest extent possible.

8 Make complete provision for replacement. If it is necessary to use a unit with a relatively high failure rate, it must be possible to replace it easily and rapidly. This does not necessarily increase reliability, per se, but it may decrease possible damage to other components during replacement. It does make an inherently less expensive system.

9 Know when to derate and to what extent. Know when to use redundancy and standby units and to what extent.

Specifying standard items increases reliability. It uses readily available materials and components of proven reliability, based on experience, sound manufacturing, good quality control, and installation techniques. Spare items are readily available. The standard items, obviously, must be suitable for the intended use.

Standardization can be extended by the building-block technique of combining standard subsystems and assemblies in various ways. This allows use of experience with the building blocks while permitting more specialized design in the total system.

Never stop proof-testing, no matter how good past experience has been with a vendor or standard items.

Remember that no change in design or procedure is (necessarily) "insignificant."

A designer must be competent in his own area of specialty. He must also be familiar with all other technology bearing on his problem—or—he must seek assistance from other specialists.

10-8 DESIGN CHECKLIST

A systematic methodology is very helpful in many situations. Design for reliability is no exception. The following list of items, as a checklist of sorts, may be useful:

1 Product requirements: (a) Are all functional, reliability, and other requirements carefully and clearly specified? (b) What are the requirements on environment? Are they reasonable? Are they based on experience or conjecture? (c) Are the reliability requirements too rigorous, too loose, or inconsistent?

2 Preliminary design: (a) Are proven designs available which can fill functional requirements? (b) Are standard units and devices available for the purpose? (c) Is the environment sufficiently different to change the performance of standard units? (d) Is environmental extrapolation necessary? If so, how much? (e) Is expert advice needed? Is it available? Will you use it?

3 Design analysis: (a) What is the behavior of each material and component under the stated environment? (b) How dependable are the available data? (c) Are available data sufficient to allow calculation of reliability? If gaps exist, can they be filled? (d) Can complete breadboard and/or prototype units be built for testing? (e) What are the weak points in the design? Have all of them been found? (f) Does the reliability meet the requirements or is redesign in order?

4 Corrective action: (a) Is expert advice needed? Will it help? (b) Are some answers available from manufacturing or quality control? (c) Is reliability limited by a very few components? If so, can they be derated, redesigned, or eliminated? Is redundancy the answer? Is a standby configuration the answer? (d) Is it possible to change the environment, e.g., cooling, heating, shielding, shock mounting, etc? (e) Is redesign in order?

(It may be necessary to cycle through steps 3 and 4 a number of times.)

5 Final design: (a) Can production, inspection, and purchasing departments assist in writing complete and clear specifications? (b) Is 100 percent inspection necessary? If so, can specifications be written to ensure this? (c) If all characteristics of the components cannot be checked, do suitable manufacturing and quality control procedures exist to minimize possible unsuitability in the components? (d) Which components will be purchased or subcontracted? Is there a list of qualified and approved vendors? (e) Is it possible to inspect and test in such a manner that substandard items will be rejected early in the manufacturing stage? (f) What is the minimum number of tests and inspections needed at each stage? What confidence level is associated with these minimum numbers? Is it necessary to test for all characteristics? (*g*) How much burn-in, debugging, etc. are required to eliminate early-life failures in equipment sent to customers? (*h*) Is a "shakedown" test of whole system necessary? (*i*) How much testing can be done without a sacrifice in product life? Is this too little, about right, or too much?

6 *Redesign after pilot operation:* Return to step 3 and proceed again!

10-9 SPECIFICATIONS AND TOLERANCES

The basic purpose of a specification is to provide an absolute, positive identification of what is desired. Ideally, specifications should be brief, concise, and simple without loss of completeness. They should also be unambiguous, self-contained, and enforceable. Reliability specifications normally have explicit definition of (1) purposes, (2) mutual understanding and agreement, (3) possibility of achievement, and (4) time of completion. Specifications are frequently written by engineers under the assumption that they are communicating directly with other engineers who will understand what is desired. Instead, specifications are often channeled through administrators (including accountants and laywers) of one company to those of a second company and thence to the engineers. Obviously, the intended and actual communications are not necessarily the same. Contracts and specifications describe not only design requirements but also explicit requirements for the entire reliability program including processing, fabricating, assembly, storage, and shipping.

Distinction should be made between specification limits and natural tolerance limits. Specification limits are set (perhaps somewhat arbitrarily) by the designer, often without regard to what can be achieved in a particular process. The natural tolerance limits are the actual capabilities of the process and can be considered as the limits within which all but a given allowable fraction of the items produced will fall. If the item produced has a normal distribution, a good design will usually have a mean value (μ) coincident with the nominal value and a standard deviation (σ) which permits only a small fraction of the items to fall outside of the specification limits. In other words, the natural tolerances will coincide with the specification limits. In any event, for a process to be acceptable, the natural tolerances must fall *within* the specification limits. For normal distributions, the natural tolerances are very often taken as $\pm 3\sigma$, since this includes all but 0.27 percent of the items.

10-10 HUMAN FACTORS

A number of systems require operation by a human, who thereby becomes an integral part of the system, and, as such, can have a significant effect on system reliability. Obviously, then, the designer must consider a large number of human factors. For example, he must not schedule the operator to (1) exceed his physical strength, (2) perform too many functions simultaneously, (3) perceive and process more information than is feasible, (4) perform meticulous tasks under difficult environmental conditions, (5) work at peak performance for long periods, (6) work with tools in cramped quarters, etc. The aim of the designer is to maintain maximum system reliability by keeping the operation as simple as possible.

Yet no matter how well a product or system is designed and constructed, its ultimate performance is subject to human factors, e.g., carelessness. One may fail to torque and secure a fastener properly or one may fail to install an explosive device properly. It is important for the designer to remember that any *detail,* no matter how minor, can cause failure. Attention to detail results, to a very significant degree, in high inherent and demonstrated reliability. This may well require complete and painstaking administrative control.

10-11 DESIGN REVIEW

Although reliability should be continuously monitored during the entire period of design, development, and test, periodic design reviews should be conducted. These are to:

1 Review progress of the design.
2 Monitor reliability growth.
3 Assure that reliability requirements will be met.
4 Provide feedback of information to all concerned.

Design review is an effort, through group examination and discussion, to ensure that a product (and its components) will meet reliability requirements. In any design of some complexity, there is necessity for a minimum of three design reviews: conceptual, interim, and final. Conceptual design reviews have a major impact on the design with successive interim and final reviews having relatively less effect as the design becomes more fixed and less time is available for major design changes. Design reviews should include all aspects of producibility and maintainability.

A design review is conducted by a design review board (or group) composed of mechanical design engineers, electrical design engineers, reliability engineers, packaging engineers, consultants as required, management representatives, etc., and chaired by the design manager or his representative. Vendor participation in conceptual and final reviews is highly desirable. A design review checklist should be prepared by the reliability engineer well in advance of the actual design review meeting. This checklist should be thoroughly detailed in covering all aspects of the design and expected performance including parts lists, part application (including possible multiple functions), tolerance studies, environmental considerations, drift and aging characteristics, maintenance factors, tradeoff studies (if any), redundancy, test data, etc.

If properly conducted, a design review can contribute substantially to avoiding serious problems by "getting the job done right the first time." Formal design reviews are effective barriers to "quick and dirty" designs based on intuition without adequate analysis.

Obviously, a large variety of approaches can be used. Except for simple series systems, it is desirable (and usually necessary) to prepare a logic (or block) diagram (Sec. 4-6). If there is any degree of complexity, a failure-modes-and-effects analysis (FMEA) is essentially required. A complete FMEA includes every component, all possible modes of failure for each component, all effects on the component and on the system for each failure mode, probability of occurrence of each failure mode, all causes for each failure mode, possible means of prevention and/or correction of each failure cause, and any other pertinent information. Tabular forms of delineating this information are often useful. Tree diagrams are also a useful graphical representation which is especially valuable when failure modes (or submodes) are not independent. The tree diagram is a plane figure having a finite number of line segments (branches), with each submode or sequence in turn represented by the appropriate number of additional line segments. Each event and its probability is shown in its proper relative position on the total tree diagram.

10-12 HOW MUCH RELIABILITY?

The question of how much reliability a given component or system requires is similar to the question of adequate performance (Sec. 1-3). There is no one answer, since each system will have its own specifics which cannot be anticipated in any general all-inclusive formula. The answer to the question is a matter of value judgment necessitating tradeoffs involving a large number of design factors such as functional purpose, time, weight, space, maintenance, human safety, legal considerations (i.e., products liability aspects), and economic considerations. *The answer to the question of how much reliability comes from a value judgment based on consideration of the complex interactions of all factors involved in the specific system.* Such determinations are obviously not easy to make.

Reliability requirements, goals, and objectives are sometimes established by the customer, sometimes by the producer. Various governmental agencies such as the Department of Defense (DOD) and the National Aeronautics and Space Agency (NASA) normally are quite definite on required reliability in purchase specifications, as are large industrial customers, such as airlines and automotive industries. Alternatively, for products to be supplied to a number of small users, the producer normally determines reliability requirements on the bases of customer needs and competitive practices. The producer has a dilemma in that he can price himself out of the market by setting and maintaining too high reliability standards, but he can also get into serious trouble by too low standards.

Design requirements, whether determined by customer or producer, should never be regarded as immutably fixed. Rather, they should be periodically and carefully reviewed and challenged with regard to need and validity of any and all design requirements (including reliability). If an unnecessarily unrealistic requirement can be removed in the early stages of a design program, substantial time, effort, and money can be saved and possibly translated into concentration on other design problems, resulting in a substantially better product.

If the final system is complex, it is necessary to break the overall reliability requirement into a series of subrequirements for the various elements of the system. The lowest reliability requirements possible should be given to design areas of great complexity or to those which require the largest advancement in the state of the art, while design areas which are simple and/or use well-known and well-tested principles or components should have the most rigorous reliability requirements. The specifics of apportioning reliability are often challenged by design supervisors and/or by the design review board. As design progresses, adjustments of reliability requirements are often made on the basis of relative progress in various design areas. Specific reliability limitations may be imposed on the designer (hopefully after proper review and consideration of all factors). For example, the designer may be prohibited from using specific materials, parts, or practices such as natural or synthetic elastomers with undesirable properties, or he/she may be required to use specific materials, practices, or parts such as solid-state electronic devices (transistors, diodes, etc.) rather than tubes.

In cases where human life is involved, it is customary to design for long lifetimes with a low probability of failure. (What is acceptable as a long lifetime and as a low probability of failure is still subject to definition in each specific case!) Where this design is for buildings and other structures for public use, the requirements and techniques to be used are often established by legal regulations, i.e., building codes. These, in turn, are subject to challenge, but a successful challenge is much more difficult than negotiating with a private customer.

Since about 1960, there has been a sharp increase in products liability cases. This activity is based on concern for safety and protection against injury of the "man on the street" by whom products are both used and misused with little or no understanding of how the product really works or potential consequences of misuse. The underlying philosophy on the part of the plaintiff attorneys is that the general public has a right to protection from use of products which may result in injury (or death) to the user. In general, the courts have held that such a right is appropriate. This obviously places a burden on the designer to make the product more reliable, both in the sense defined in Chapter 1 and in the sense that the public understands it.

The producer has the duty to design and manufacture a product which is not only safe for intended use but also safe in all foreseeable, although unintended, uses. Anticipation of possible misuse, perhaps unusual but not unlikely, is an important part of the designer's task. It is a real exercise in creative, uninhibited, imaginative thinking, and the designer should seek as many independent views and suggestions as possible.

Where human life or safety is clearly not involved, design is normally made for minimum total cost while meeting functional and reliability requirements. This total cost involves purchase cost, installation cost, cost of labor and parts for replacement, shutdown expenses during repair or replacement, and cost of production losses.

Economic factors are always important elements in any design. One commonly seeks to obtain a minimum cost consistent with functional design objectives and requirements. Obviously, this is not always easy to accomplish. One technique which can be very helpful is known as value engineering (or value analysis). While this technique can be valuable in review procedures and in redesign of products, it is even more valuable if employed from the start of a design.

The first and most important step in value analysis is definition of function, i.e., exactly what the part is supposed to do and exactly what is required of the part. One then determines which requirements are important and which, although desirable, are not truly needed. Assessment is made for any important secondary functions such as appearance, convenience in use, noise levels, etc. Elimination of unnecessary functions can be very important. If done properly, one can establish a chart delineating functions and relating subrequirements to individual subassemblies or units. This permits exposure of functional overlap and often permits elimination of subunits.

Once the functional requirement has reached an irreducible minimum, the next step is to explore and search for alternative approaches to accomplish the same function. This requires flexibility of thinking and creativity on the part of the designer. This requires not only much more than finding a different source of a component of the same type, but also requires consideration of new configurations and/or procedures which will fulfill the functional requirement as well as, if not better than, the original. Ideas that are not sufficient in themselves can be useful in suggesting combinations which may provide working solutions.

Once the alternatives have been established, cost analyses are made. If the entire process is completed in the early stages of design before the design is released, determinations can be made of whether it is more economical to make or buy components and assemblies, if some parts can be eliminated, and if alternative materials, designs, and processing techniques can be used. Cost analyses should be broken into basic, step-by-step, recurring, and nonrecurring costs. Costs must be authentic and obtained from reliable sources so that cost comparisons are realistic and not "crystal-ball" estimates.

For example, one company had seam welded electronic chassis for several years to provide radio-frequency shielding since "everyone knew" that spot welding did not provide suitable shielding. This was questioned during a value analysis. Samples using both welding techniques were prepared and tested. No appreciable difference in shielding was found. As a result the company changed to spot welding which saved about $2000 for each chassis type in a product having more than 100 chassis types.

Another organization was specifying "overseas packaging" for computer punch cards even though they were being used entirely within the continental United States. This specification had been established because there had been occasional damage in shipping using "domestic packaging" but even less with "overseas packaging." The difference in cost of 5 cents per thousand cards may seem insignificant until one realizes that the total order was 1.5×10^9 cards. One could certainly absorb some damage within the $75,000 lower cost using domestic packaging.

How much reliability? There is no one answer which fits all situations. Each system and its application must be examined on its own merits with all interested and responsible parties being involved in the decision as to required reliability.

10-13 SUMMARY COMMENT

It must be recognized that this chapter does not really tell how to design for reliability. It is intended to provide an introduction to some methodology, to indicate the myriad number of aspects that must be considered, and to give some feeling for the complexity of the task of designing for reliability.

PROBLEMS

10-1 Consider the system shown in the figure, in which there are four relays in the circuit. If the probability of an open-circuit failure in any one relay is 0.02, what is the probability of open-circuit failure of the system? Is this a complete model? If not, what is missing? How would the more complete model look in terms of the reliability block diagram? What would the reliability of the system be?

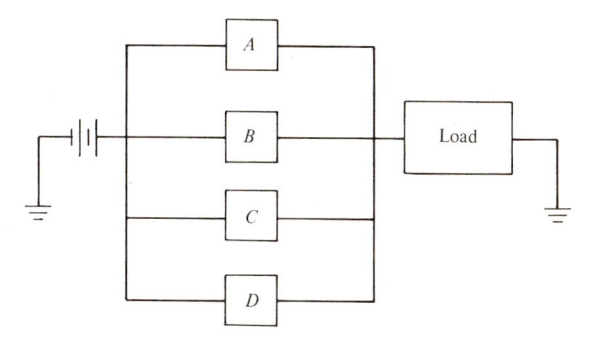

10-2 Given the failure rates below, in failures per million hours of operation, for Prob. 4-14, what are the minimum, mean, and maximum reliabilities for the system given for 1000 h of operation? What is the practical significance of these calculated reliabilities?

Unit	λ, 10^{-6} failures/h		
	Min.	Mean	Max.
A	4	8	15
B	4	10	20
C	24	30	36
D	5	10	15
E	35	40	50

10-3 The estimated failure probability for an element that can fail short-circuited or open-circuited is 0.14. The ratio of short-circuited to open-circuited failure probabilities is known to be about 0.04. What is the optimum number of parallel elements to use? What is the optimum number of series elements? If short circuits are more likely to occur, e.g., $q_0/q_s = 0.50$, what is the optimum design?

10-4 The mechanical aspects of an electrical motor are being examined in the hope of offering a 500-h guarantee. The motor has two brushes, two ball bearings and a small fan on the shaft which draws air around the field coils. Electrical failures (overload, burnout, and insulation breakdown) are random with an MTTF of 20,000 h. Brush failure is normally distributed, with a mean of 1000 h and a standard deviation of 200 h. Bearing failure is also normally distributed, with a mean of 1800 h and a standard deviation of 600 h. Unpredictable mechanical failures (abrasive dust on the bearings or commutators, etc.) have a useful-life MTTF of 10,000 h. Would you advise the sale of this motor with a guaranteed life of 500 h? If not, what life would you be willing to recommend for guarantee? If this is too small, what would you do?

10-5 Consider a relatively complex system which can be maintained reasonably well, e.g., a shipboard radar, in which a few components (although the best available with current technology) have MTBFs on the order of 10–50 h. A system of high reliability is desired but obviously cannot be obtained with these relatively unreliable components. Should the rest of the system be built with equally unreliable components or should the rest of the system be built with highly reliable components? Under what conditions, if any, might one view be appropriate? Under what conditions, if any, might the other view be appropriate? Elaborate (briefly) on your answer.

10-6 The most used and most complex component in a radio control system is the semiconductor. Transistors, diodes, and integrated circuits are the heart of most any electronic equipment produced today. There are so many manufacturers and an endless selection of devices to choose from, it is nearly impossible to select the one device that has all the right specifications to suit the design, particularly when there is a requirement for very high-reliability. The most obvious choice is a thoroughly screened and qualified device or exhaustive testing, right? Elaborate (*briefly*) on your answer.

10-7 Consider various ways in which human factors can influence reliability and maintainability. Comment *briefly* on two or more of these and how they exert their influence.

PART TWO

11
A POSSIBLE TIMESAVER

11-1 THE PROBLEM

In concept, determination of system reliability is simple. Only four steps are required: (1) identify each possible state of the system (in this context, state is defined by listing all components and whether each is functioning successfully or has failed); (2) determine those states which result in successful operation of the system; (3) calculate the probability of each successful state occurring; and (4) sum these probabilities to obtain the system reliability. (This essentially summarizes Chapter 7.)

If we apply this concept to the system shown in Fig. 11-1 with the assumption that each component has only two modes of operation, i.e., successful function or failure with a reliability r_i and an unreliability q_i, then we discover that this particular system with ten components has $2^{10} = 1024$ states. A more complex system will have a larger number of states. Doubling the number of components to twenty results in 2^{20} states, i.e., nearly 1.05×10^6 states.

Obviously, the task of direct computation of reliability for even a moderately complex system can be overwhelming when using hand calculations or even when using a computer. How, then, can the reliability of a complex system be obtained while expending only a reasonable amount of time and effort?

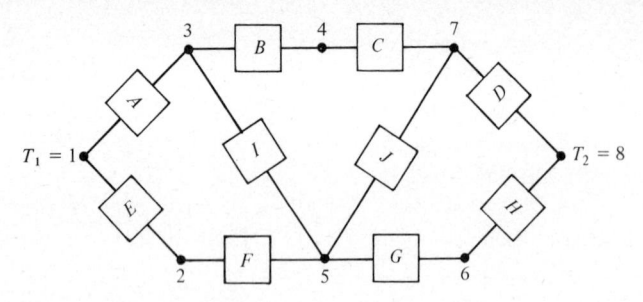

FIGURE 11-1
A network system with ten components.

11-2 ONE SOLUTION

A very common and very acceptable engineering approach to a complex situation (like the one indicated in the preceding section) is to make a model or an approximation which can be solved much more readily than the actual problem. While this reduces the effort required to get a solution, it also raises questions as to the validity of the approximation and how close the answer obtained approaches the answer in the real situation. Obviously, one would like to get an "exact" answer through the approximation. Next best is an answer which is close to the exact answer and on the conservative side. In this context, such an approximation should underestimate rather than overestimate the reliability of the given system.

For a network (e.g., Fig. 11-1) which has two terminals with all components interconnected at nodes in the network, there is an approximation technique, known as the minimal-cut approximation, which reduces the computational procedure and gives a good lower bound to the total system reliability. The closeness of the estimate of reliability provided by this method depends on the arrangement of the specific network. While it is difficult to estimate the error in general, the approximation is close when the reliabilities of the individual components are close to unity. The method outlined below applies specifically to components which are bidirectional, but it can be extended to unidirectional components. The method is founded on the concepts of minimal cuts and coherent systems.

A coherent system must fulfill four conditions: (1) when a group of components in the system fails and thus causes the system to fail, the occurrence of any additional failure (or failures) will not restore the system to a functional condition; (2) when a group of components functions properly and thus the system functions properly, the return of some failed components to proper functioning will not cause the system to fail; (3) when all components in the system function properly, the system functions properly; and (4) when all components in the system fail, the system fails.

A cut is a set or group of components such that if all components in the group fail, the system fails independently of the performance of the other components in the system. A system may have a large number of cuts and any given component may be in more than one cut.

A minimal cut is a cut in which there is no specific subset of components whose failure alone will cause the system to fail. This concept is vital since the approximation to system reliability depends upon identifying all the minimal cuts in the system. The lower-bound approximation for system reliability is the probability that there is failure in none of the minimal cuts in the system. This is not a true reliability expression since there is an implicit assumption that failures of minimal cuts occur independently. This is not always true since one (or more) component may be in more than one minimal cut.

This approximation can be expressed in general terms if the jth minimal cut is considered to be the set S_j. The component members of this jth minimal cut are given by $i \in S_j$. The probability of failure in this jth minimal cut is

$$Q_j = \prod_{i \in S_j} q_i \qquad (11\text{-}1)$$

The reliability approximation for the system is given by the probability that none of the minimal cuts in the system fail, i.e.,

$$R_{Mc} = \prod_{\text{all } j} \left[1 - \prod_{i \in S_j} q_i \right] \qquad (11\text{-}2)$$

A cut is a set of components such that when the components in the cut are removed from the system there is no path from one terminal to the other. Removal of the components in a minimal cut separates the system into exactly two subsystems. One subsystem contains one terminal with a path from that terminal to every other node in the subsystem. The other subsystem contains the second terminal which has a path to every other node in the second subsystem. Figure 11-2 shows an example of a minimal cut in Fig. 11-1 and the two subsystems formed by removing this minimal cut.

The problem of identifying all the minimal cuts is the same as the problem of generating the set of two-part partitions of the set N into subsets X and $\bar{X}$. Set X must define a connected subset that includes node T_1 (terminal 1), while set $\bar{X}$ must define a connected subset that includes node T_2 (terminal 2).

To be certain that no minimal cut is overlooked, some sort of systematic procedure is necessary. This can be done by using an algorithm developed by Jensen and Bellmore (11-1). The algorithm constructs a tree which consists of vertices and edges. Figure 11-3 shows the complete minimal-cut generation tree for the system shown in Fig. 11-2. The entire set of minimal cuts for this system is given in Table 11-1. Edges are line segments in the tree and vertices are points. Each vertex is given an integer index for easy identification. The tree vertex indexed 0 is known as the root vertex. Minimal cuts appear in the tree at terminal vertices, i.e., those vertices which touch only one edge. The algorithm assures that each terminal vertex represents a unique minimal cut and that each minimal cut is represented by a terminal vertex.

Tree edges are labeled by T or F, where t or f is an integer representing a node of the system. T and F are used to indicate members of the sets X

134 INTRODUCTION TO RELIABILITY IN DESIGN

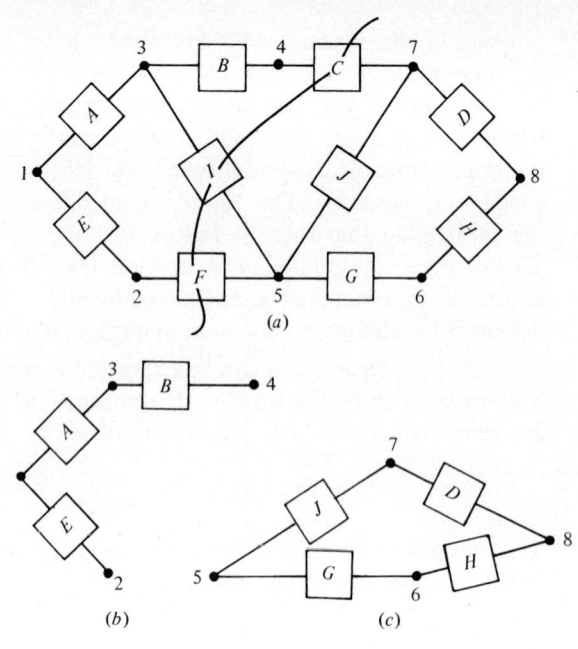

FIGURE 11-2
A minimal cut with subsets of system in Fig. 11-1; (*a*) minimal cut (*C*, *F*, 1), (*b*) subset $X = (1, 2, 3, 4)$, (*c*) Subset $\bar{X} = (5, 6, 7, 8)$.

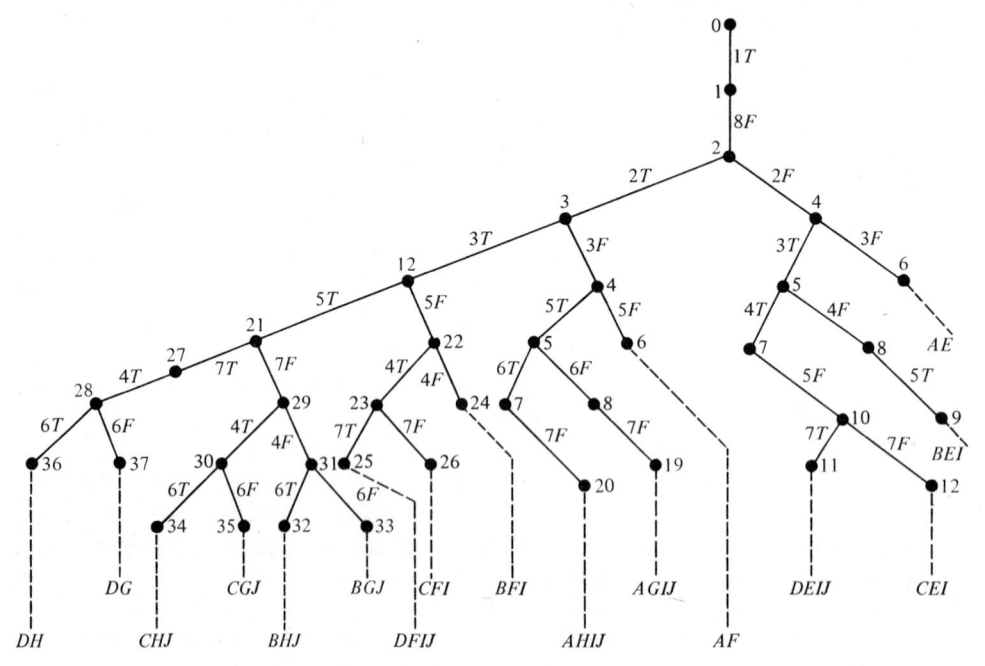

FIGURE 11-3
Complete minimal-cut generation tree for Fig. 11-1.

Table 11-1 MINIMAL CUTS OF THE SYSTEM IN FIG. 11-1, SPECIFIED BY THE SET

	Node, members of the set X								
Cut	1	2	3	4	5	6	7	8	Components in the cut
0	X								AE
1	X	X							BEI
2	X		X	X					CEI
3	X		X	X			X		DEIJ
4	X	X							AF
5	X	X			X				AGIJ
6	X	X				X	X		AHIJ
7	X	X	X						BFI
8	X	X	X	X					CFI
9	X	X	X	X			X		DFIJ
10	X	X	X		X				BGJ
11	X	X	X		X	X			BHJ
12	X	X	X	X	X				CGJ
13	X	X	X	X	X	X			CHJ
14	X	X	X	X	X		X		DG
15	X	X	X	X	X	X	X		DH

and $\bar{X}$, respectively. For example, on the path from vertex 0 to vertex 20, labels 1T, 2T, 5T, and 6T appear on edges. These identify the set X for the minimal cut represented by vertex 20. Table 11-1 shows the correspondence between this subset X and the cut $AHIJ$. Labels 8F, 3F, and 7F also appear on edges on this same path. These three nodes are in subset $\bar{X}$. In general, the nodes corresponding to the edges labeled with xF do not constitute the entire $\bar{X}$ subset. The subset $\bar{X}$ for a terminal vertex is found by subtracting subset X from the total set.

The generation tree is sequentially constructed from a starting point of no edges and no vertices. Edges and vertices are developed by the algorithm. Each time a terminal vertex is developed, a minimal cut has been found and its contribution to the unreliability is calculated. Labels for edges are chosen in such a manner that the subsets X and $\bar{X}$ define connected subsystems for each terminal vertex. The enumerative nature of the tree assures that every minimal cut is generated and that no minimal cut is generated twice.

11-2-1 The Algorithm

The following algorithm determines the set of all minimal cuts of a system between nodes T_1 and T_2.

1 Create three vertices for the tree indexed 0, 1, and 2, and edges (0,1) and (1,2) labeled tF and fF, respectively. Let vertices 0 and 1 be scanned and vertex 2 be unscanned. Vertex 0 is called the root vertex. Go to step 2.

2 Choose the unscanned vertex with the greatest index and mark it scanned. If there are no unscanned vertices, the algorithm terminates, for the complete tree has been generated. The vertex chosen will be denoted as vertex i.

Find the unique simple path 1_i that connects the root vertex with vertex i. Identify the sets Y_{1i}, Y_{2i}, Y_{3i}, and W_i as defined above. Choose y, an element of the set W_i. If W_i has no members, go to step 7.

Construct the subnetwork defined by the set of nodes $Y_4 (= Y_{2i} \cup Y_{3i} - y)$. Test to see if it is connected. If not, go to step 4. If so, go to step 3.

3 Create two new vertices indexed k and $k + 1$ where k is one greater than the number of vertices currently in the tree. Vertices k and $k + 1$ are unscanned. Create two new edges (i,k) and $(i,k + 1)$, labeled yT and yF, respectively. Go to step 2.

4 The subnetwork defined by Y_4 is not connected. Find the set of nodes Y_5 that defines the connected subnetwork that includes node t. If $Y_{2i} \subset Y_5$, go to step 5. If $Y_{2i} \not\subset Y_5$, go to step 6.

5 Create vertex k and edge (i,k) labeled yT where k is one greater than the number of vertices currently in the tree. Determine the set $Y_6 = Y_4 - Y_5$. For each number $Z \in Y_{6i}$, create a vertex of the tree and an edge labeled zT. If $[Y_6]$ is the number of members in the set Y_6, vertices $k + 1, k + 2, \ldots, k + [Y_6]$ will be created. Edges $(k, k + 1), (k + 1, k + 2), \ldots, (k + [Y_6] - 1, k + [Y_6])$ will also be created. Finally, create vertex $k + [Y_6] + 1$ and edge $(i, k + [Y_6] + 1)$ labeled yF. Go to step 2.

6 Create one new vertex indexed k and an edge (i,k) labeled yF. Go to step 2.

7 A minimal cut has been generated at this step. The set X_i for the cut is $X_i = Y_{1i}$ and $\bar{X}_i = N - Y_{1i}$. The components in the minimal cut are those that have one terminal in the set X and one in the set $\bar{X}$. Let these components be the set S_j. Find the probability of failure of the minimal cut:

$$q_{Sj} = \prod_{i \in S_j} q_j$$

Include this in the system unreliability estimate:

$$Q_i = Q_{i-1} + q_{Sj} - Q_{i-1}(q_{Sj})$$

where Q_{i-1} is the unreliability estimate before the discovery of this cut. To generate more minimal cuts, return to step 2.

The algorithm is designed to generate the entire tree and maintain it in the core memory of the computer. In the computer program implementation of the algorithm, a great deal of space in the core memory of the computer must be set aside to keep all the information concerning the vertices, edges, and labels of the tree. A very simple modification to the algorithm is possible that at any point in the generation process allows one to keep only that portion of the tree necessary to discover the set of minimal cuts that have not yet been generated. The modification takes place in the first paragraph of step 2.

2 Choose the unscanned vertex with the greatest index. Let this be *i*. Discard those vertices in the tree with indices greater than *i*. Mark vertex *i* scanned. If there are. . . .

The algorithm then continues as before.

With this change, the process of cut generation is practically limited only by the computer time one is willing to expend, while without this change the process is limited by the core size of the computer.

EXAMPLE 11-1 Consider Prob. 4-32. Determine the reliability of this system using the minimal-cut approximation and compare with the exact reliability.

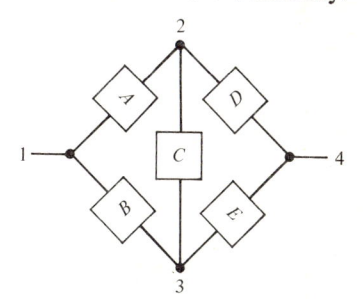

In this system there are $2^5 = 32$ states. There are sixteen cuts as indicated in Table 11-2. Comparison of this table with the figure allows easy recognition of the minimal cuts which are listed in Table 11-3.

Table 11-2 CUTS OF THE SYSTEM BETWEEN NODES 1 AND 4

Cut	Components in cut
1	A,B
2	A,B,C
3	A,B,D
4	A,B,E
5	A,B,C,D
6	A,B,C,E
7	A,B,D,E
8	A,B,C,D,E
9	D,E
10	C,D,E
11	B,D,E
12	A,D,E
13	B,C,D,E
14	A,C,D,E
15	A,C,E
16	B,C,D

Table 11-3 MINIMAL CUTS BETWEEN NODES 1 AND 4

Minimal cut	Components in minimal cut	Probability of failure of minimal cut
1	A,B	$q_A q_B$
2	D,E	$q_E q_D$
3	A,C,E	$q_A q_C q_E$
4	B,C,D	$q_B q_C q_D$

If we assume that all components are identical, then the exact reliability of this system is given by

$$R_{sys} = r^5 + 5r^4 q + 8r^3 q^2 + 2r^2 q^3$$

From Eq. (11-2), the approximate reliability from the minimal-cut method is

$$R_{Mc} = (1 - q^2)^2 (1 - q^3)^3$$

Comparisons of the numerical evaluation of these two equations are given in Table 11-4 for three different values of reliability of individual components.

Table 11-4 COMPARISON OF RELIABILITY CALCULATIONS

Component		System reliability		
Ω	$q = 1 - r$	Exact	Minimal cut	Difference
0.90	0.10	0.9784800	0.9781408	0.0003392
0.95	0.05	0.994780625	0.994757514	0.000023111
0.99	0.01	0.9997980498	0.9997980104	0.000000394

Even for this relatively simple system, it is obvious that some effort is saved by using the approximation rather than the exact solution. It is also obvious that the difference between the two answers is small and that this difference decreases as the reliability of the individual components increases. ////

12
FAILURE RATES AND DESIGN

12-1 THE PROBLEM

Determination of reliability with a good degree of accuracy can be difficult in a number of circumstances. In electronic equipment, in particular, this can be a major problem, since failure-rate data may become very limited as the quality of components improves. For example, demonstration of failure rates of the order of one failure in 10^8–10^{10} h of mission time in laboratory life tests is very difficult and not economically feasible.

How would you obtain failure-rate data to use in making designs? What influence would you expect differences in calculated and observed failure rates to have on design? How does prior experience influence new design?

12-2 FAILURE-RATE ESTIMATION

One possible source of usable data in lieu of actual failure-rate data is field replacement-rate data. Tanner (12-2) studied four military systems projects to obtain replacement data of various electronic components. These four systems used over 5.5×10^6 electronic components which had a total accumulated operating time of over 82×10^9 h. Some of these systems, but not all, operated under controlled temperature and humidity conditions. All components were of MIL-type specifications (or better) and operated at stress and power levels well below rating in analog and digital computer type circuits.

Experience showed an average replacement of $1\frac{1}{2}$ components per failure incident. A number of these failures were caused by secondary failures, handling, testing, maintenance, etc. In this context, a replacement is "any circumstance that causes the system to degrade sufficiently to require that component parts be replaced after the system has been installed and checked out." The replacement rates of the various components in the four systems (labeled A, B, C, and D) are given in Tables 12-1 through 12-5. In general, replacement rates for resistors and capacitors were of the order of 0.1 to 5 per 10^9 h while the rates for semiconductors were of the order of 5 to 20 per 10^9 h.

Examination of each replaced component gave no evidence of wearout failure, but showed that cause of failure was invariably related to a manufacturing defect, abuse, or overstressing. In system C, replacement of semiconductors was 20 times greater during the first quarter of operation than during the last quarter. This was due to early system operation while operators were still un-

Table 12-1 REPLACEMENT RATES OF RESISTORS

Resistor	Sys.	Electrical stress (%)	No. used (thousands)	Component h (billion)	Replacements	Replacement rate/10^9 h Avg.	80% Confid. Lts.	
Fixed								
Carbon comp.	B	20	14.8	0.192	0	5.2[a]	0	12.0
	C	25	184.7	2.290	10	4.4	2.7	6.8
	D	25	125.3	1.960	11	5.6	3.6	8.5
Carbon film	A	20	299.9	8.850	18	2.0	1.4	2.8
	B	10	2420.0	31.400	13	0.4	0.3	0.6
	D	10	20.1	0.314	2	6.4	1.7	17.0
Metal film	C	5	22.0	0.277	3	11.0	4.0	25.0
Power WW	B	15	53.8	0.700	2	2.9	0.8	7.7
	C	60	5.0	0.064	1	16.0	1.7	62.0
	D	50	0.9	0.014	0	71.0[a]	0	163.0
		Totals:	3146.5	46.06	60			
Variable								
Carbon	D	Low	1.9	0.029	0	34[a]	0	78
WW	B	Low	1.5	0.019	3	155	57	346
	C	Low	1.6	0.019	0	53[a]	0	122
		Totals:	5.0	0.067	3			

[a] Assumes one replacement.

Table 12-2 REPLACEMENT RATES OF CAPACITORS (FIXED)

Capacitor	Sys.	Electrical stress (%)	No. used (thousands)	Component h (billion)	Replacements	Replacement rate/10^9 h Avg.	80% Confid. Lts.	
Dielectric								
Ceramic	D	6	38.8	0.610	48	79.0	65.0	96.0
Glass	B	5	715.0	9.300	1	0.1	0.01	0.4
	C	7	9.4	0.124	0	8.1[a]	0	19.0
Mica	A	5	115.1	3.430	0	0.3[a]	0	0.7
	D	5	28.1	0.440	2	4.6	1.2	12.0
Paper	C	30	1.7	0.020	0	50.0[a]	0	115.0
	D	45	0.9	0.014	2	143.0	38.0	380.0
Plastic	B	10	6.3	0.082	0	12.0[a]	0	28.0
	D	5	1.0	0.016	0	44.0[a]	0	100.0
		Totals:	916.3	14.04	53			
Electrolytic								
Aluminum	D	30	2.7	0.043	7	164.0	91.0	275.0
Tantalum solid	B	60	3.3	0.043	0	23.0[a]	0	53.0
	C	60	16.5	0.210	0	4.8[a]	0	11.0
	D	45	3.7	0.058	1	17.0	2.0	66.0
Tantalum wet	A	65	6.1	0.180	6	33.0	17.0	58.0
	D	45	7.0	0.110	5	45.0	22.0	83.0
		Totals:	39.3	0.644	19			

[a] Assumes one replacement.

Table 12-3 REPLACEMENT RATES OF MAGNETIC COMPONENTS

Component	Sys.	Electrical stress (%)	No. used (thousands)	Component h (billion)	Replacements	Replacement rate/10^9 h Avg.	80% Confid. Lts.	
Coil, RF	B	Nom.	10.0	0.130	0	8[a]	0	18
	C	Nom.	2.1	0.026	1	38	4	148
	D	Nom.	11.9	0.186	7	38	21	64
Reactor, power	D	Nom.	0.1	0.022	0	476[a]	0	1095
Transformer								
Power	D	Nom.	0.7	0.011	1	91	10	354
Audio	D	Nom.	0.2	0.003	0	333[a]	0	766
RF	D	Nom.	1.2	0.019	0	53[a]	0	122
Pulse	C	Nom.	7.8	0.098	7	71	39	119
		Totals:	34.0	0.495	16			
Relay, armature	C	50	0.5	0.006	1	182	19	708
	D		1.4	0.022	14	637	430	917
		Totals:	1.9	0.028	15			

[a] Assume one replacement.

Table 12-4 REPLACEMENT RATES OF DIODES

Diode	Sys.	Electrical stress (%)	No. used (thousands)	Component h (billion)	Replacements	Replacement rate/10^9 h Avg.	80% Confid. Lts.	
Germanium								
Low power	D	1	2.8	0.044	3	69	25	154
Silicon								
Low power	B	5	51.5	0.670	7	10	6	17
	C	5	119.3	1.400	56	40	33	48
	D	1	11.8	0.185	2	11	3	29
Med. power	A	10	100.4	2.970	27	9	7	12
	C	6	45.0	0.580	29	50	39	64
	D	40	5.8	0.090	2	22	6	59
Zener, med	D	30	10.7	0.167	5	30	15	56
		Totals:	347.3	6.10	131			

familiar with the system rather than to early failure in the components (Sec. 10-10).

It is apparent from the data that replacement rates differed considerably among the four systems. These differences can be explained by differences in design, component quality, packaging techniques, stress levels, workmanship, operating environment, etc.

As an indication of the effect of improving reliability, the transatlantic cable system, placed in service on 25 September 1956, may be of interest.

Table 12-5 REPLACEMENT RATES OF TRANSISTORS

Transistor	Sys.	Electrical stress (%)	No. used (thousands)	Component h (billion)	Replacements	Replacement rate/10^9 h Avg.	80% Confid. Lts.	
Germanium								
Low power	B	25	907.0	11.800	187	16	14	18
	D	15	11.1	0.170	17	98	69	136
Silicon								
Low power	A	10	61.2	1.850	68	37	31	43
	D	17	6.2	0.097	14	144	97	207
Med. power	C	5	51.3	0.670	61	91	76	107
	D	35	1.6	0.026	0	39[a]	0	90
High power	C	—	1.0	0.013	30	2290	1770	2930
	D	20	0.1	0.002	2	925	246	2460
		Totals:	1039.5	14.63	379			

[a] Assumes one replacement.

This system links 102 repeaters (amplifiers), containing 306 electron tubes and 6600 passive electronic components, located on the ocean floor where maintenance is impossible and replacement extremely expensive. The reliability goal was a maximum of one failure in 20 years of service. A failure in any one electron tube would cause system failure, since there was no redundancy: the design was based on the concept of a minimum number of components each having the utmost reliability.

Additional cables were subsequently laid and put into operation. As of 1 August 1965, a total of 3.87×10^9 component hours of operation had accumulated without a single component failure, thus surpassing the reliability design goal.. Performance figures are given in Table 12-6.

Achievement of this reliability goal required some relatively unusual procedures. Only components with long records of trouble-free performance were even considered. Premium materials manufactured under rigid specifications were procured. Extreme precautions were taken to prevent degradation or contamination of these materials during assembly. All components were tested under extreme conditions to detect early failures. Design specialists were present at all times during manufacture to challenge anything which could affect the ultimate reliability. Great attention was focused on inspection.

This example provides an excellent illustration of the point that reliability cannot be "tested or inspected into" a product or system. It must be designed

Table 12-6 SUBMARINE CABLE PASSIVE COMPONENTS

Component	Electrical stress (%)	No. used (thousands)	Component[b] h (billion)	Replacements	Replacement rate/10^9 h Avg.	80% Confid. Lts.	
Resistors							
Carbon comp.	10	12.7	0.262	0			
Power WW	10	2.4	0.049	0			
Precision WW	10	40.4	0.837	0			
	Totals:	55.5	1.148	0	0.87[a]	0	2.0
Capacitors							
Mica	10	42.5	0.879	0			
Paper	40	27.4	0.567	0			
Polystyrene	5	9.5	0.197	0			
	Totals:	79.4	1.643	0	0.61[a]	0	1.4
Magnetic							
IF&RF inductors	Nom.	46.2	0.958	0			
IF&RF transformers	Nom.	5.9	0.121	0			
	Totals:	52.1	1.079	0	0.93[a]	0	2.1
All components	Grand Totals:	186.9	3.87	0		0	0.6

[a] Assumes one failure.
[b] Component hours computed to August 1, 1965.

and manufactured into the system. An additional factor was the great emphasis on training of personnel and close surveillance during manufacture to minimize opportunity for human error. A suggested lesson we might consider is that designers should strive to design equipment which is as easy to operate and maintain as possible, in order to minimize human error. This is obviously more desirable than having some one looking over your shoulder.

12-3 CALCULATED VERSUS ACTUAL FAILURE RATES

Techniques and procedures for predicting reliability can be found in many places. Many require combining individual component failure rates, adjusted for specific design stresses and environments, to obtain subsystem and system reliability estimates.

While the predicted reliability number is often of value by itself, it also has value (combined with estimates of performance, weight, and cost) as a basis

Table 12-7 CONFIGURATIONS AND FAILURE RATES: A COMPARISON

Type of part	Calculated failure rate (λ)	Configuration 1 Number of parts (n)	($\lambda \times n$)	Configuration 2 Number of parts (n)	($\lambda \times n$)
A	0.5	70	35	190	95
B	0.01	500	5	200	2
C	1.2	85	102	100	120
D	0.05	135	7	120	6
E	0.1	10	1	0	0
F	0.2	5	1	5	1
Total		805	151	725	225

Type of part	Actual failure rate (λ)	Configuration 1 Number of parts (n)	($\lambda \times n$)	Configuration 2 Number of parts (n)	($\lambda \times n$)
A	0.7	70	49	190	133
B	0.85	500	425	200	170
C	1.2	85	102	100	120
D	0.8	135	108	120	96
E	2.0	10	20	0	0
F	1.0	5	5	5	5
Total		805	709	725	524

for tradeoffs and improvements at the component level. Obviously, wrong decisions can be made on tradeoffs if reliability calculations are not accurate.

Consider a hypothetical example, Table 12-7, in which a given function can be accomplished with either of two possible configurations. On the basis of the calculated failure rate (individual laboratory failure rates, adjusted for stress and environment), configuration 1 with a total rate of 151 (compared with 225 for configuration 2) should be selected if all other things are equal. Actual field failure rates, however, might change things so that configuration 2 with a total failure rate of 524 (compared with 709 for configuration 1) should be selected. Thus, while it is possible to obtain different reliabilities for different sets of failure rates, it is also possible to be misled into deciding which is most reliable. While this kind of situation may not happen very often, it is impossible to know *when* it does happen. Estimates of failure rates can provide valuable information, but they can also be misleading as a basis of comparison of alternative configurations.

Differences between predicted failure rates and observed failure rates depend on many factors, such as:

1 Definition of failure.
2 Actual environment compared with prediction environment.
3 Maintainability, support, testing equipment, and special personnel.
4 Composition of components and component-failure rates assumed in making the prediction.
5 Manufacturing processes including inspection and quality control.
6 Distributions of times to failure.
7 Independence of component failures.

12-4 PRIOR EXPERIENCE

Use of past experience to optimize a new design is a challenge. Field reliability data appear in extreme quantities and subtle forms. As a result, they are more likely to be used qualitatively than quantitatively. In addition, equipment which can supply sufficient and reliable field data has often become technically obsolescent enough to preclude direct application in a new design. Further, disparity is often found between design (calculated) and observed reliability data.

A reliability design review of the SP-50 automatic pilot (12-3) shows that it is possible for the reliability prediction based on summation of random failures to be exceeded in operation, despite the normal expectation. Various adjustment factors for system complexity are not necessarily valid for avionics systems. These factors are really allowances for inadequacies in design or deficiencies in manufacture that result in chronic failures (chronic failure being defined as a repetitive failure), whether of an obvious or subtle nature. Elimination of chronic failures is necessary if significant increases in reliability are to be obtained.

13
PROJECT MERCURY

13-1 THE SITUATION

In the late 1950s and early 1960s, Project Mercury, under NASA, had the problem of achieving manned orbital flight with a reasonable degree of reliability and safety at the earliest possible time. In this context, reliability was defined as the probability that a given mission would proceed to completion without mishap while flight safety was defined as the probability of crew survival during a given mission. Two design guidelines were paramount: (1) no single failure should cause an abort, and (2) no single failure during an abort should result in the loss of life of the crew. Three distinct, but interrelated, areas were involved: namely, launch vehicle, spacecraft, and operation.

The Redstone missile and the Atlas rocket were well developed at that stage and could potentially serve for launching. For the spacecraft, however, there was no parallel background experience. In like manner, the operations aspects had limited background experience and information.

How would you proceed, keeping in mind that one objective is to "achieve orbital operation at the earliest possible time"?

13-2 THE PROJECT MERCURY APPROACH

In this situation, a combination of quantitative and qualitative procedures were undertaken. These will be discussed below in terms of the three areas previously indicated.

13-2-1 Launch Vehicle

The Redstone missile (for suborbital flights) and the Atlas rocket (for orbital flights) were well developed in the early 1960s. Only relative minor modifications were necessary to meet the requirements of the Mercury system. A vigorous effort was initiated to retain proven components within the entire system so that the fund of knowledge and experience gained from testing and operation of the launch vehicles was not lost.

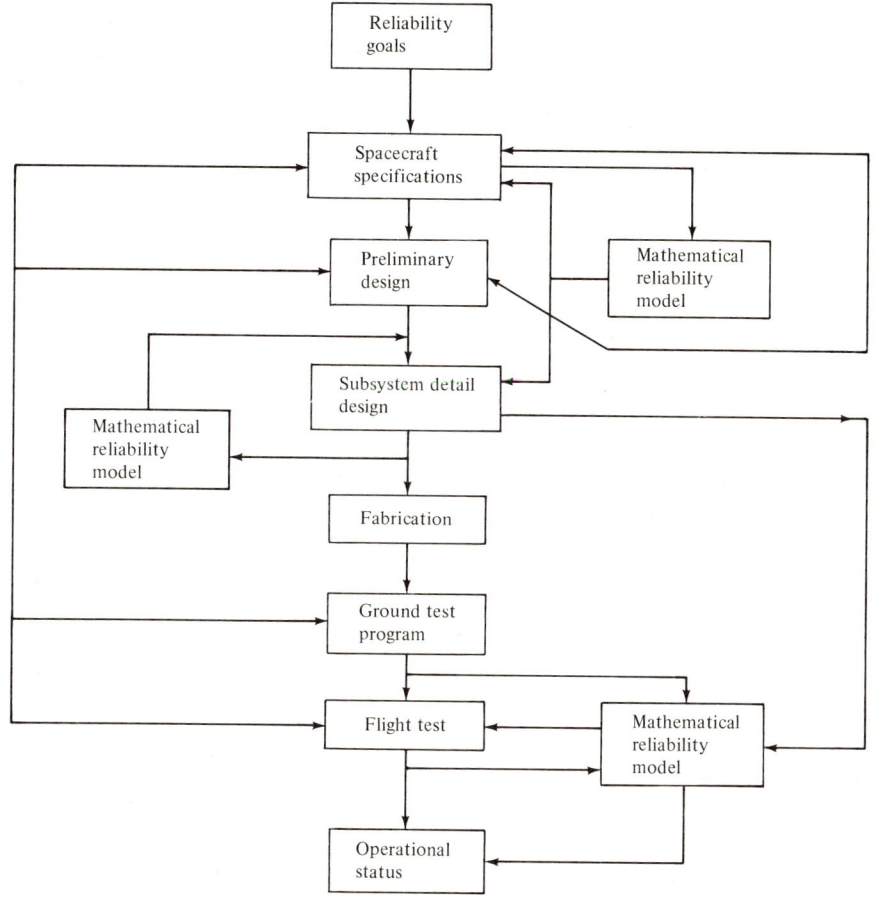

FIGURE 13-1
Quantitative reliability model.

An abort-sensing and implementation system was incorporated into the launch system. The purpose of this was to "sense impending catastrophic launch vehicle failure, automatically generate an abort command, and activate the spacecraft escape system in sufficient time to assure astronaut safety." Prior to launch, nominal values were assigned to critical launch vehicle performance parameters. If specified parameter tolerances were exceeded, a signal was generated to initiate the abort sequence. The reliability of the equipment for this operation was provided with the free use of parallel and standby redundancies.

Since the reliability problems of the launch vehicle were based on components of proven performance with system modifications accomplished through existing components, the quantitative reliability model (Fig. 13-1) was applicable. This technique uses failure-rate data to obtain numerical values for the probability of failure-free operation of the system. Analysis progresses from definition of reliability goals, through subsystem design and fabrication, to ground test, and finally to system operational status. The mathematical model is modified at various points on the path (and at various times), with revision incorporating additional data and information gained as the design progresses.

13-2-2 Spacecraft

While the quantitative approach was applicable to the launch vehicle because failure-rate data existed, there were limited similar data available for design of the spacecraft. In addition, development of such data would require extensive time and testing, as indicated in Fig. 13-2 which indicates the number of tests

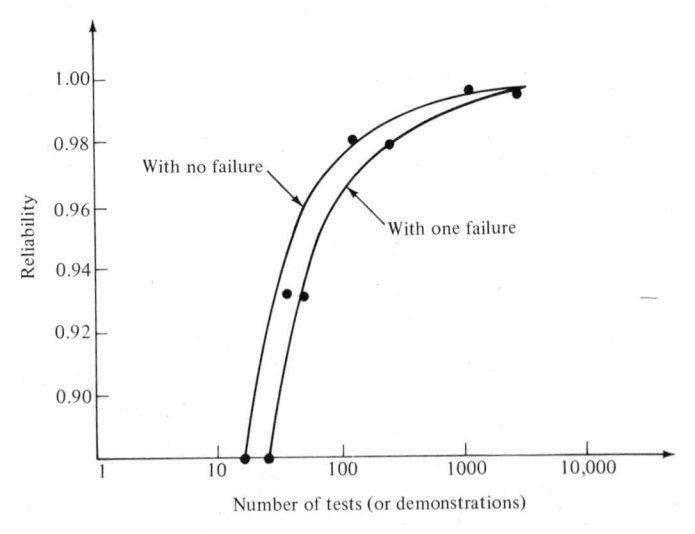

FIGURE 13-2
Number of tests required to establish a given reliability with 90 percent confidence.

required to establish a system reliability (with 90 percent confidence) if zero failures or one failure occur during the test program. In effect, to establish a reliability of 98 percent (with 90 percent confidence) for the environmental control system (which had to operate for a two-week mission time) would require 240 weeks of failureless test or 500 weeks of a "one-failure" test. Under such conditions, the "earliest possible time" using a quantiative approach would be quite remote. Obviously, an alternative approach was in order.

The approach taken could be called a qualitative reliability model. A detailed failure mode (and resulting effect) analysis (FMEA) was made of the entire system. Each element, usually at its lowest level of replacement in the assembly, was analyzed to determine all possible failure modes. The effect of

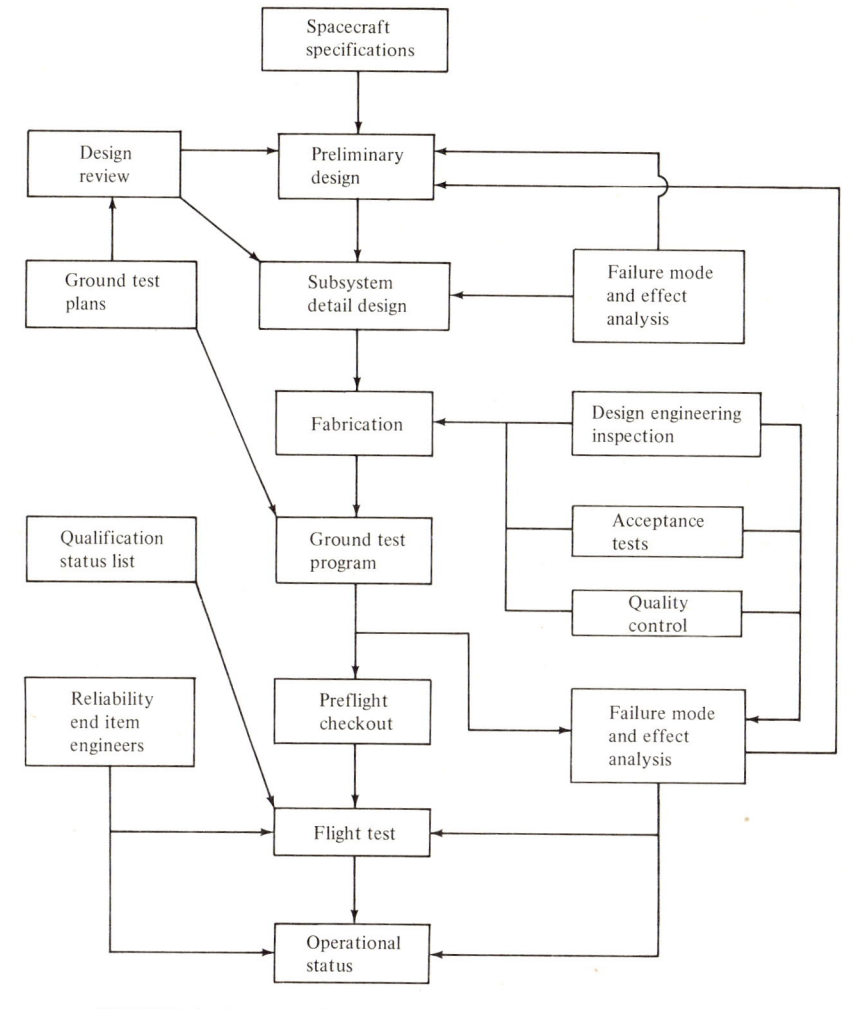

FIGURE 13-3
Quantative reliability model.

each failure mode on the subsystem and the achievement of mission objectives was determined. In determining the effects of potential failure of each element, full consideration was given to redundancy, alternative operational modes, and alternative operational procedures. If suitable alternative ways of achieving mission success and crew safety were not available, then modification was made. In effect, this qualitative approach seeks to maximize the number of functional paths within a subsystem or system. The qualitative reliability model is shown schematically in Fig. 13-3.

The quantitative model was used for elements and subsystems when failure-rate data were available. This model also served as a guide for obtaining reliability during the design stage. Flow diagrams from the qualitative model were applied to subsystems and systems operation to show sequences of action, primary and critical abort paths, crew inputs, and the principal functions of individual units. A combination of the two models permitted finding the reliability of the spacecraft.

13-2-3 Operations

In operations, the findings from the two other areas were combined into a net, composite, mission reliability. Units, previously tested as separate entities, were joined with "neighbors" to study the interface behavior of the assembly. Simulated flights, under both normal and excessive conditions, served as final tests of previously determined failure-rate and failure-mode conclusions. Operations, in effect, was a final check of the quantitative-qualitative model combination. The general features of all three areas are shown in Fig. 13-4.

13-3 COMMENTS

The use of the quantitative-qualitative model was a legitimate approach to determining system reliability for Mercury since the launch vehicle and a number of the spacecraft components were "proven" units with a substantial amount of failure data. In addition, the relative simplicity of the system (in comparison with Gemini and Apollo which followed) allowed for detail mapping of alternative "successful function" modes. At that point in the space program, the approach was an acceptable technique. It can be applied in a number of similar situations.

The Gemini and Apollo missions incorporated many new components and designs which could best be approached through the qualitative model before adequate experience data were available. Concentration on failure-mode analysis, with emphasis on multiple functioning paths, is therefore highly appropriate. A major problem is locating every possible point of failure and providing a detour around it.

Concentration on proven units has a definite advantage because of known failure data and performance characteristics. Complete reliance on such units, however, ignores the possible advantages of innovation. Introduction of new

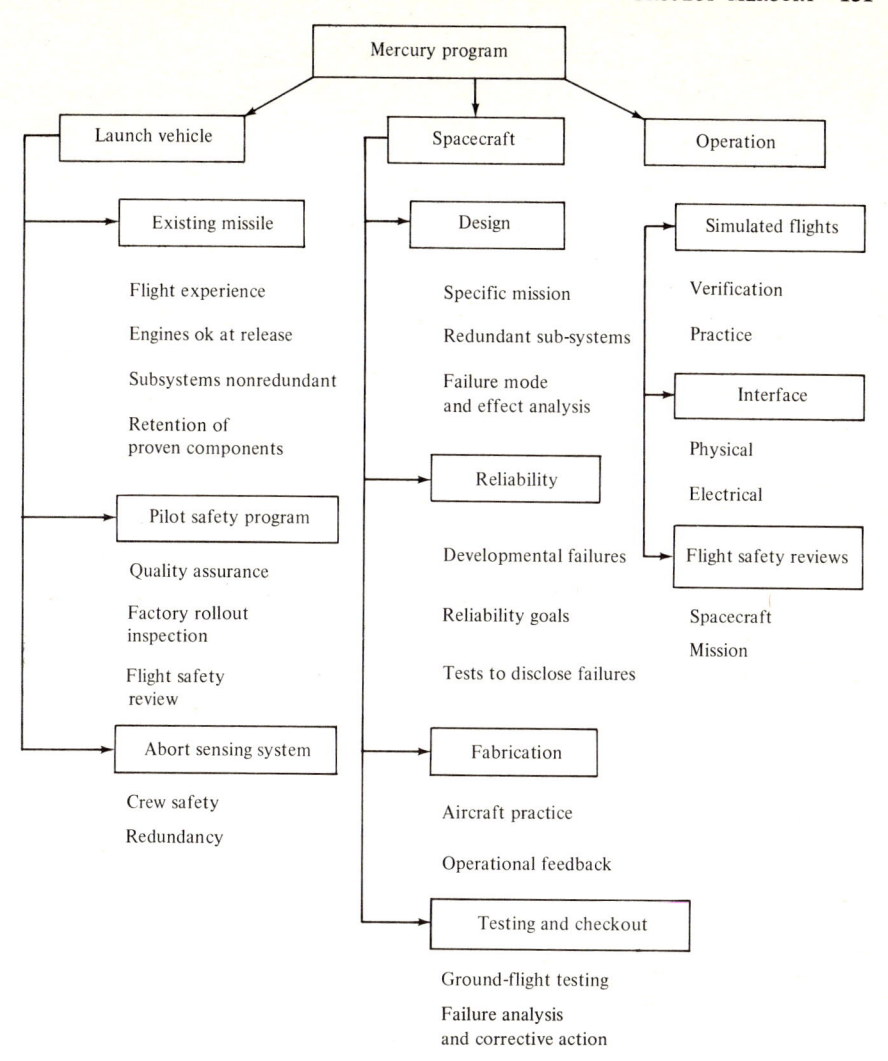

```
                          Mercury program
                                │
          ┌─────────────────────┼─────────────────────┐
     Launch vehicle         Spacecraft            Operation
          │                     │                     │
    Existing missile          Design            Simulated flights

    Flight experience      Specific mission       Verification
    Engines ok at release  Redundant sub-systems  Practice
    Subsystems nonredundant Failure mode
    Retention of           and effect analysis      Interface
    proven components
                             Reliability           Physical
    Pilot safety program                           Electrical
                          Developmental failures
    Quality assurance                           Flight safety reviews
    Factory rollout        Reliability goals
    inspection                                     Spacecraft
                          Tests to disclose failures Mission
    Flight safety
    review

    Abort sensing system   Fabrication

    Crew safety            Aircraft practice
    Redundancy
                           Operational feedback

                          Testing and checkout

                          Ground-flight testing
                          Failure analysis
                          and corrective action
```

FIGURE 13-4
Mercury operational reliability model.

types of components and refined ideas may reduce system cost, may decrease the mass-to-volume ratio, may give improved performance, and may result in a unit that is more reliable than the earlier unit which it replaced.

14

DESIGN OF A DC-DC CONVERTER

14-1 THE PROBLEM

Conditioned electrical power required for future space missions will increase into the multikilowatt range as missions become more complex and of longer duration. The accompanying need for more efficient, lighter weight and more reliable power conditioning will also increase.

In this context, a converter is needed to be used with a lightweight, high-efficiency, pulse-width-modulated inverter rated at approximately 1 kW per phase. Since the inverter was designed to operate without an output transformer, it required a regulated high-voltage dc input. A converter was required to produce 1 kW of power, with a 150 percent continuous overload rating and with a regulated 200-V ($\pm 1\%$) dc output, from an unregulated input of 56 V, $+10\%$ and -20% dc. Design, size, weight, efficiency, and reliability were considered equally important.

How would you proceed to develop such a converter, giving appropriate attention to designing reliability into the system? What would your converter configuration be?

14-2 ONE WORKING DESIGN CONCEPT

An experimental quasi-square-wave dc-dc converter was designed with a breadboard model being built and tested. A complete schematic circuit diagram with parts list is shown in Fig. 14-1. The output stage used transistors in a full-bridge arrangement which offers speed, gain, and saturation characteristics resulting in high efficiencies. Toroidal powdered-metal (Ni-17Fe-2Mo) cores (instead of laminated cores) were used in a low-loss power transformer operating at 7-kHz carrier frequency. An integrated-circuit control section and a proportional-current drive system were part of the total system.

The design thus features high-frequency operation, minimum number of components, elimination of critical components, a minimum of adjustments, component derating, low-power integrated circuits, circuits which cannot produce potentially damaging voltage or current spikes, and an overload-protection circuit.

A primary advantage of high-frequency power conversion is the ability to reduce size and weight because of the smaller magnetic components. High-frequency operation, however, results in increased core and switching losses, so that a compromise is necessary to obtain acceptable weight and efficiency.

A breadboard model was built and tested, using bench power supplies and a load bank. Separate power supplies were used to power the control and voltage drive circuits. Overall efficiency measurements were accurate within 1 percent.

The overall efficiency of the converter as a function of input voltage and output power is shown in Fig. 14-2. The peak measured efficiency was over 92 percent at a 600-W output with a 56-V input. The efficiency was 88 percent at 1 kW. It was greater than 87 percent from 150 W to 1.2 kW, and was almost independent of the input voltage.

The entire system had less than 100 parts with a total weight of about 2.7 kg (6 lb) in the breadboard configuration. No size or weight estimates were made for a packaged system. If the carrier frequency were increased, lower weight would be possible but at the price of a lower efficiency. These parts could be optimized for a specific mission; the result would be as much as 25 percent weight reduction with some loss in efficiency.

Reliability is critical in all aerospace equipment. One approach to gain reliability is incorporation of protective devices throughout the system. This, however, results in increased complexity with a need for more critical parts. This specific converter, with less than 100 parts, requires no selected, matched, or precision components. Only three variable resistors are required if R_4 (Fig. 14-1) is used to set the output voltage precisely.

The control circuit utilized several integrated circuits. Most of the discrete components could be replaced by hybrid thin- or thick-film circuits. Power dissipation and peak electrical stress in the control and voltage-drive sections are very low, thus contributing to long-life expectancy. Voltages, currents, and powers in the current-drive and output stages are much higher, but the component ratings are at least 150 percent of the peak electrical stresses.

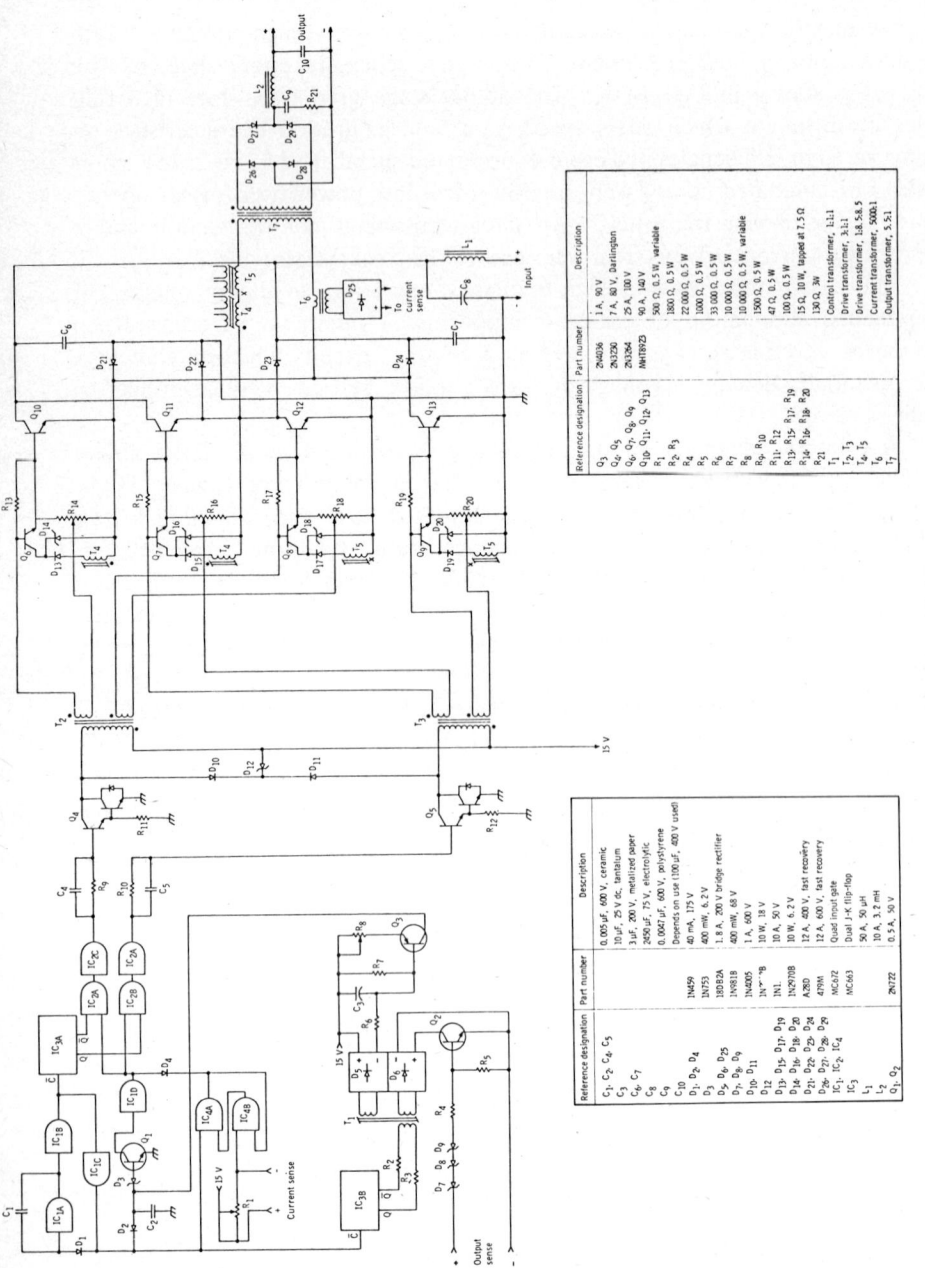

FIGURE 14-1 Converter circuit diagram and parts list.

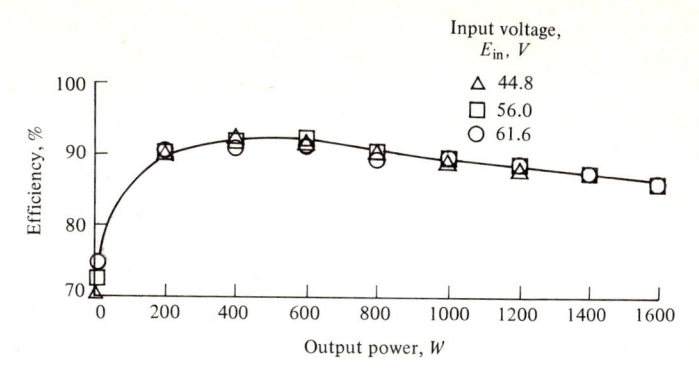

FIGURE 14-2
Converter efficiency as a function of output power; output voltage = 200 V.

The configuration of the full-bridge output circuit is inherently reliable (e.g., the voltage across the transistors is reduced by a factor of 2). The penalty paid for this is the required large number of power transistors, each one having a collector-emitter drop and base-drive loss. A two-transistor circuit would measurably reduce these losses. The transistors required for this arrangement, however, would require a higher voltage than the voltage actually used. In addition, their gain and saturation characteristics are not comparable, and the net result would be cancellation of most of the possible advantages. The power transformer for the two-transistor circuit would also have to be larger (for equal efficiency); this would result in more severe voltage-spike problems and thus effectively reduce reliability.

The bridge circuit used requires only a very simple transformer with no split or tapped windings. This is an advantage. Winding configuration is important because stray capacitance and inductance must be minimized for efficient and reliable operation.

15
AVAILABILITY CONCEPT

15-1 THE PROBLEM

As space missions increase in numbers and length, problems in reliability and crew safety increase to the point where a "catastrophic" failure is quite probable, e.g., on a Venusian or Martian mission. A lunar mission (e.g., Apollo) takes about a week. A Martian mission, in contrast, will take 500 to 700 days with a further complication that, once on the trans-Mars path, there is no practical abort short of the planned mission profile.

Table 15-1 gives an indication of the reliability which might be required and the reliability available on the basis of contemporary state of the art. The useful-life model was used, taking into account normal redundancy and actual duty cycles of equipment. The predictions are obviously unreasonably low for long-term missions. If the effective MTBF is increased by adding redundancy, a point of diminishing returns is soon reached since system complexity increases at an even greater rate because of additional functions such as monitoring and switching.

Increased reliability can be expected from developments and improvements in technology. Experience shows, however, that in a five-year period, this increase will be a factor of 5 to 10 (Fig. 15-1). It is obvious from Fig. 15-2

Table 15-1 MISSION REQUIREMENTS AND RELIABILITY

Mission objective	Duration (h)	Reliability Required	Reliability State-of-art estimate
Lunar Lander	200	0.95	0.95
Venus Flyby	8,000	0.95	0.03
Mars Flyby	16,800	0.95	0.006

that an increase of more than one order of magnitude is necessary to obtain the desired reliability for an interplanetary mission, whether in the near or more distant future.

Estimates have been made that as few as 6 failures might be encountered on a Martian mission. It should be noted that as many as 85 might also be expected. It seems obvious that failures will occur in various subsystems. These, in turn, can lead to total system failure.

If failure is defined as any catastrophic event which results in immediate or eventual loss of a crew member, how would you handle the problem of designing a Martian flyby system with the desired reliability?

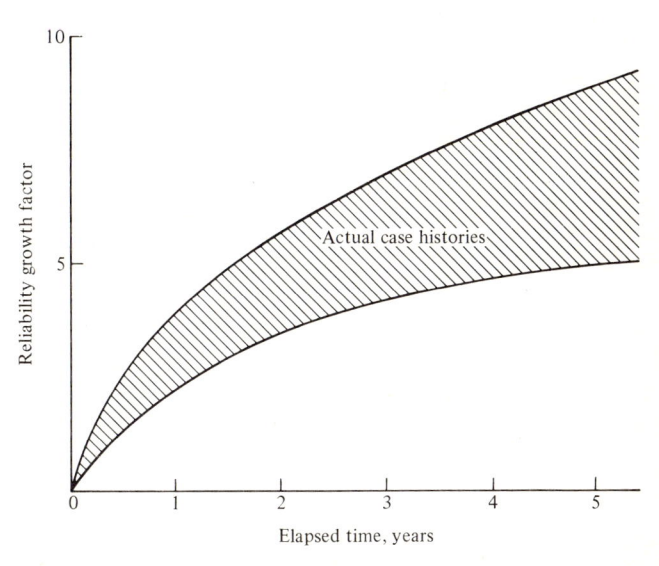

FIGURE 15-1
Growth in reliability as a function of time.

158 INTRODUCTION TO RELIABILITY IN DESIGN

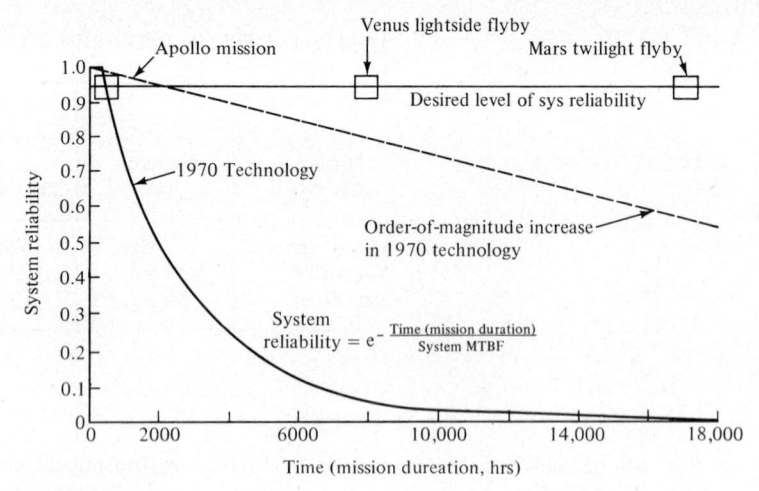

FIGURE 15-2
Interplanetary flight reliability as a function of mission duration.

15-2 ONE SOLUTION

Recognizing the "certainty" of subsystem failures, provision should be made to compensate for temporary loss of function and to correct the malfunction with minimum risk to the crew. This can be approached by applying the "availability concept," i.e., a design or mission-analysis technique that seeks the specific man-machine relationship which will maximize mission effectiveness by establishing a safe and reasonable balance among system performance, reliability, and maintainability. Solution is obtained by identifying potential failures, contingencies, and constraints associated with them, and using these as design and operational criteria.

In the foregoing context, the potential design reliability may, at times, be somewhat compromised in order to meet safety-imposed maintenance-time constraints. This is essentially a question of high probability of no failure in comparison with a high probability of in-flight maintenance, since the two are not always compatible. Consider the case of a pump (Table 15-2) which can be

Table 15-2 RELIABILITY AND MAINTAINABILITY FOR A PUMP INSTALLATION

Method of installation	Failure rate per 10^6 h	Maintenance time required
Welded	0.002	1 h
AN fittings	6.8	30 min
Quick disconnects	13.5	5 min

connected into the system in at least three ways. Each method has its own reliability and maintenance characteristics. In this example, the highest reliability has the lowest maintainability and vice versa. This is typical of the multitude of problems which must be resolved. Spare-weight allowance constrains design criteria on one extreme, while downtime constrains them on the other extreme.

System availability is defined as that fraction of the total mission time during which the system is completely operable. There is an implied assumption that maintenance can be performed which will restore a function to "good as new." In a space vehicle, this includes all logistics times, such as failure-isolation time, delay in obtaining parts, etc. These times are of major concern in space travel although they may be of relatively minor importance in other situations. Availability (A) is thus the probability of functioning at any specific time in the mission profile, i.e.,

$$A = \text{operable time/mission time} \qquad (15\text{-}1)$$

$$A = \frac{\text{MTBF}}{\text{MTBF} + \text{MTTR}} \qquad (15\text{-}2)$$

where MTBF = mean time between failures
MTTR = mean time to repair

Availability expresses the probability of being completely operable at any given time. It may be apportioned or allocated in the same general way as reliability. Differences in weighting criteria, however, may be different because of crew-safety requirements.

A number of factors place severe limitations on availability. These factors, which must be very carefully determined and applied, are (1) downtime restrictions, owing to backup system, crew limitations, etc.; (2) weight constraints of spares, tools, and test equipment; (3) astronaut maintenance capability and limitations; (4) necessity of undertaking a specific repair action at a given point in time; (5) monitor and fault-isolation capabilities and limitations; and (6) nonrepairable systems (or functions within systems).

Evaluation of expected downtime per failure relative to downtime in a subsystem function is necessary to obtain a proper decision on mission availability limitations and contingencies. Typical examples are given in Table 15-3. The allowable downtime per failure depends on the contribution to crew survival and mission usage time. This may be limited by use of a backup system or by crew tolerance to a degraded mode of operation. Obviously, the crew should not be exposed beyond an allowable risk level. Total downtime must include fault isolation, obtaining spares, and active repair time.

The availability of any given subsystem can be calculated from

$$A_{ss} = \exp(-\lambda T e^{-ut}) \qquad (15\text{-}3)$$

where λ = equipment failure rate = 1/MTBF
T = mission time
u = repair rate = 1/MTTR
t = maintenance time constant (MTC)

Table 15-3 AVAILABILITY LIMITATIONS

Subsystem	MTC Allowable downtime, h/incident	MTTR Maximum anticipated downtime (90%), h[a]	Limitations or contingencies
Structure	2	Internal 0.4	1 in. puncture or separation
		External 1.0	System failure
Life systems	2	0.5	No circulation thermal control or suit operations
Control systems	8	0.4	No functions—nonspinning limiting case
Power system	8		Battery operation or bypass loops and minimum loads
		Isotope system 2	
		Fuel or solar cells 1	
	1	0.5	Processing or distribution out
Guidance and navigation	Over 24	0.4	No power maneuvers possible
Communication and data	24	0.3	All functions—very remote
Instrumentation	Over 24	0.7	Critical sensor assumed redundant

[a] Ninety percent of more probable failure modes.

For a situation in which the equipment MTBF = 1600 h, T = 10,080 h, MTTR = 0.4 h, and MTC = 8 h, Eq. (15-3) gives A_{ss} = 0.999999987. In this case, the probability of making repairs is very high, with only 23 chances in 10^9 of not being able to effect repairs. The expected average number of repairs can be found by dividing T by MTTF to give 6.3 per mission for this specific problem.

In a real sense, the effect of successful repairs is equivalent to standby redundancy. The repair is instantaneous, so far as system and mission performance is concerned, if it is accomplished within the MTC. The number of spares can be determined from the cumulative Poisson distribution [Eq. (4-14)],

$$P = \sum_{i=0}^{i=n} \frac{e^{-\lambda T}(\lambda T)^i}{i!}$$

where P = probability of subsystem mission success
λ = equipment failure rate
T = mission time
n = number of spares required

Spares must be carried so that repairs can be made. The number of spares to be carried must be compatible with the objective of minimizing weight and/or volume. At the same time, the number of spares carried has a direct effect on the level of crew safety. The relationship between the probability of mission success (safe return) and the spares carried is shown in Fig. 15-3

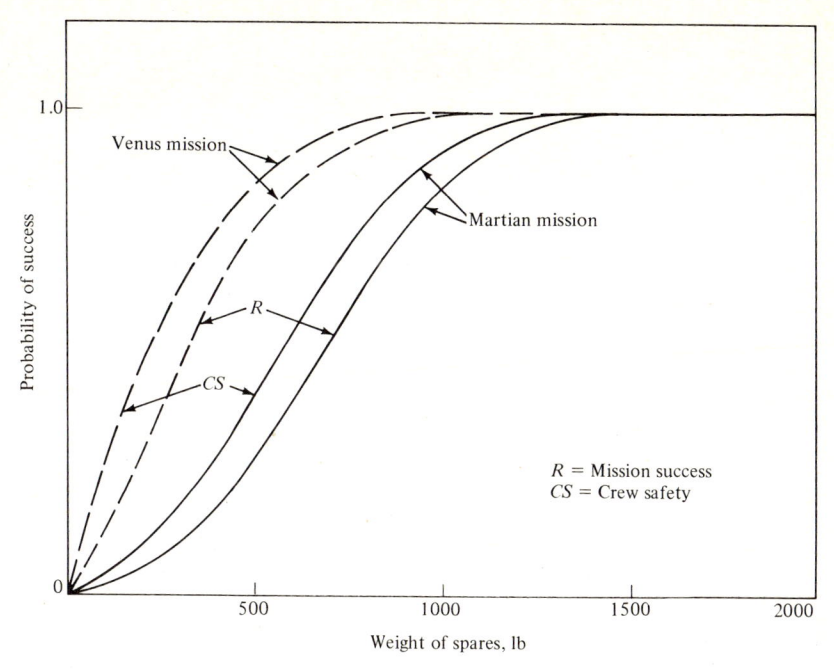

FIGURE 15-3
Probability of success as a function of total weight of spares.

for a typical Venusian mission of 360 days and a typical Martian mission of 700 days. The major difference is due to the difference in mission length. Difference between mission success and crew safety is due to service of auxiliary equipment such as instrumentation, a TV system, etc. For a typical flyby mission to Mars, a total of 185 spares with a weight of about 900 lb would be necessary to ensure reasonable crew safety. With these spares and appropriate preparations for this specific set of repairs, the probability of crew survival is predicted to be between 0.9900 and 0.9994.

Design of spares for short replacement time obviously decreases MTTR. Further reduction in MTTR can be obtained by adequate training. There is evidence that a zero-g environment, per se, does not appreciably slow a trained astronaut. Constraints imposed by a spacesuit, however, may double the task time.

Design for long-term space flights cannot depend on high reliability alone, but must be optimized around the availability concept which requires establishing an appropriate balance among system performance, reliability, maintenance complexity, and spares weight/volume. The concept is vital in space flight but is useful in many other applications.

16
HUMAN RELIABILITY

16-1 THE PROBLEM

Evaluation and prediction of potential human error is a most perplexing, pervasive, and evasive problem. Systems reliability cannot be assessed by consideration of hardware alone. It is recognized that man and his equipment are imperfect and subject to unexpected deviation from their intended function. An ability has been developed to quantitatively evaluate hardware deviation (and possibly apply control or correction thereto), but development of a similar capability relative to the human component has lagged far behind.

Relatively little is known about predicting potential human error. Still less is known about predicting performance or reliability of the man/machine interface. Human errors are continually occurring. In a number of military weapons systems, failures attributed to human error range from 12 to 60 percent.

There is a clear necessity to engineer human-factor aspects concurrently with the other aspects of a developing system, rather than wait for the laborious (and often costly trial-and-error) test and evaluation solutions. This has been accentuated by the increasing criticality of the consequence of error in present systems and by the extension of humanity's capabilities and limitations significantly beyond any previous experience.

Much is known about the human tolerance to physical stress, about human intellectual capabilities, and about human ability to transform information into an observable event. Little, however, is known about how to translate such information into terms which allow human capabilities and limitations to be integrated into practical engineering solutions.

How would you deal with this problem of "translation" in designing a reliable total system?

16-2 ONE SOLUTION

At least one technique has been developed which provides a mathematical relationship between a quantified observation of one small segment of human performance (in terms of reliability) and a predicted value of an analogous segment of performance, using similar hardware configurations in a specific situation. Through combination of observed and predicted events, it is possible to predict the reliability with which a human will perform a task and, ultimately, the contribution of the human element to the total system and its operation.

The procedure involves six steps. Initially, reliability estimates are derived for separate and discrete task elements for which no valid reliability values exist. These estimates are combined to form a total task-reliability estimate which, in turn, may be applied in system analysis. The steps are:

1 Identify the tasks to be performed: Tasks are identified at a gross level, i.e., each task being composed of a series of task elements which must be performed sequentially to accomplish the task. Each task represents a complete operation such as, "check ignition system."

2 Identify the task elements: Each major task in step 1 must be subdivided into the basic task elements (individual, discrete, simple operations) necessary to complete the task. Elements might be, "read indicator," and similar operations. Each task element involves a small segment of human performance that can be assessed in terms of error potential. In addition, such a task element may occur in a number of tasks which appear to be different. This similarity at the task-element level allows possible adoption of the scheme in a wide variety of fields.

3 Empirical performance data: Empirical performance data are necessary for at least some of the task elements to be rated. These data should be based on performance of a task element which is similar in content and hardware configuration to the task element to be rated and must contain a reliability estimate of the interaction between man and machine. These empirical data should reflect rating for conditions under which the task element will actually be performed.

Empirical data can come from in-house or field operations performed

164 INTRODUCTION TO RELIABILITY IN DESIGN

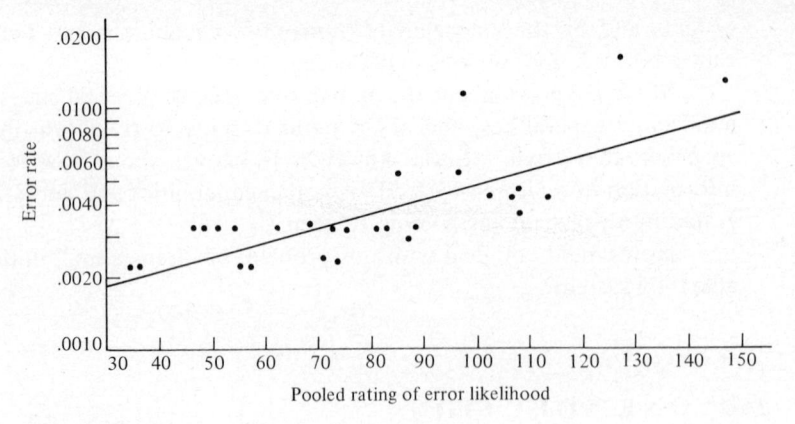

FIGURE 16-1
Relationship between rate of error and likelihood of error.

for evaluation of hardware and procedures. The methods of obtaining such data are critical in order to ensure objective reporting of error. Data may also be obtained from literature on experiments. The latter are normally results obtained by trained observers under controlled conditions. Differences in these conditions and the conditions under which the data will be applied must be accounted for.

(In the specific case reported, reliability estimates from the Payne-Altman Index of Electronic Equipment Operability Data Store were used. These data were applicable to 29 of 60 task elements being rated when extrapolated to field conditions and thus were useful for the intended purpose.)

4 *Establish task-element rating:* To obtain task-element reliability, each task element is rated in keeping with its level of difficulty, i.e., error potential. *Rating,* in this context, is a judgment of error potential applied by personnel familiar with the gross task requirements, the system or components on which the task element is performed, and the level of skill of the technician performing the task. (Any room here for human error?)

A ten-point scale, ranging from least to most error, is adequate for statistical purposes. Type of error is not important. It is sufficient that there might be an error because of task difficulty. Each task element is rated in keeping with its judged error potential. Ratings are statistically summarized, and a pooled rating is assigned to each task element being evaluated.

5 *Develop regression equation:* In order to predict task-element reliability, a relationship must be found between the empirical data and the judged ratings in the form of a regression line or an equation. This relationship should also be tested for goodness of fit. The regression line, or equation, once established, is used to provide reliability estimates for task elements for which there are no data.

In the specific study used, the data took the form shown in Fig. 16-1.

Table 16-1 TYPICAL TASK ELEMENTS: RATING MEANS AND STANDARD DEVIATIONS; RELIABILITY ESTIMATES

Task element	Rating Mean	Standard deviation	Reliability estimate
Read technical instructions	8.3	2.2	0.9918
Read time (brush recorder)	8.2	2.1	0.9921
Read time (watch)	4.1	2.1	0.9983
Read electrical or flow meter	7.0	2.8	0.9945
Read pressure gauge	5.4	2.2	0.9969
Inspect for loose bolts and clamps	6.4	1.9	0.9955
Remove lockwire	2.7	1.5	0.9993
Install lockwire	6.0	2.3	0.9961
Remove reducing adapter	3.0	1.7	0.9991
Install reducing adapter	4.9	1.6	0.9975
Loosen nuts, bolts, and plugs	2.8	1.3	0.9992
Install nuts, plugs, and bolts	4.6	1.7	0.9979
Tighten nuts, bolts, and plugs	5.3	2.6	0.9970
Apply gasket cement	5.3	2.3	0.9971
Lubricate O ring	4.5	2.5	0.9979
Fill sump with oil	4.3	1.6	0.9981
Verify component removed or installed	3.5	2.4	0.9988
Inspect for air bubbles (leak check)	5.0	2.2	0.9974

In this case, the line providing the best fit is expressed as

$$\log E = -2.9174 + 0.006122 PR \quad (16\text{-}1)$$

where E = error rate (error rate = 1 − empirical reliability)

PR = pooled rating of error likelihood

Some typical task-element reliabilities from this study are shown in Table 16-1.

6 *Establish task reliability:* Individual task elements must be identified with specific tasks. Within the task, each task element has a reliability estimate assigned as derived from the regression line. Since the task is essentially a series configuration, the total task reliability can be predicted by taking the product of the task-element reliabilities. Task reliabilities thus determined can be combined in the appropriate manner to estimate the human contribution to component and system reliability.

16-2-1 Extension of the Method

As given, the foregoing method applies to the evaluation of performance of one individual acting alone. In cases where operator or technician backup is anticipated for some, or all, task elements, this task redundancy should be included.

In the case of redundant hardware, where backup is available at all times, the resulting reliability is (as for parallel redundancy)

$$R = 1 - (1 - R_1)^n \qquad (4\text{-}6)$$

where R_1 = reliability of one system
n = number of backup systems

For task redundancy, this must be modified to account for the amount of time a second technician is available as backup for the first technician. The equation then becomes

$$R = \frac{1 - (1 - R_1)^n(T_1) + R_1 T_2}{T_1 + T_2} \qquad (16\text{-}2)$$

where R_1 = reliability of one man
T_1 = time (percent) a second man is available for redundant effort
T_2 = time remaining in which second man is available (100 percent $- T_1$)

In cases where a task or task element is critical and desired reliability improvements cannot be achieved through hardware design, or the characteristics of the situation are such that personnel backup is desirable, the foregoing method might merit serious consideration.

16-2-2 Validation

Performance data obtained during category II testing of the Titan II propulsion system were used for validation. Observed performance data were converted to performance reliability on the basis of the ratio of the number of successes to attempts. For the various task elements, the number of attempts ranged from 12 to 2821. Since confidence level is a function of sample size, the top ten task elements were selected for examination, after the data had been reviewed. A rank-order correlation and a chi-square test were performed to assess the relationship between predicted and observed reliability estimates and thus test the adequacy of the predictions. The results indicated a significant relationship. Additional validation is highly desirable but the evidence does indicate a potential usefulness for the method.

The method developed relates to the derivation of task-element reliabilities where none existed previously. Use may be limited, in terms of accurate prediction, by the validity of the empirical data and the quality of the judges providing task-element ratings.

17
SUPER-RELIABILITY

17-1 THE PROBLEM

In many situations, a reliability of 90 percent is entirely satisfactory. In other cases, 99 percent reliability is required. In a number of applications, 99.99 percent might be considered very good, while a reliability of 99.999999 percent might be regarded as phenomenally good. In a relatively slow computer performing one million operations per second, however, a reliability of 99.999999 percent implies an average of one error every 100 s. If computers did indeed make errors that frequently, it is most doubtful that they would be trusted to the degree they are in so many areas; i.e., so many people seem to regard them as essentially infallible. A reliability of 99.999999 percent, however, is reasonable (although perhaps a bit optimistic) for many electronic components, e.g., integrated circuits.

If failure is defined as an error in storing or sending a signal, how would you design a computer for a reliability of the order of 99.99999999999999 percent when the components involved have a reliability of the order of 99.999999 percent (a difference of eight orders of magnitude)?

17-2 POSSIBLE SOLUTIONS

It is obvious that highly reliable computers are being built and from components which are several orders of magnitude less reliable. Even though improvements in the state of the art can improve the reliability of individual components (and thus the computer), this fact clearly does not apply in this situation.

One obvious approach is the use of redundancy techniques. For computers, these can be regarded in two groups: signal redundancy and hardware redundancy. Signal redundancy is achieved by adding redundant information to signals that are stored or manipulated within the machine. Hardware redundancy is achieved by adding redundant circuits to process signals without changing the structure of the signals. Word length is not increased, but each word is processed by more circuits. (The following discussion treats hardware redundancy only, since the concepts may be more broadly applicable than those of signal redundancy.)

Redundancy can be applied at various levels, with the level of application greatly influencing vulnerability to multiple failure. As an example, consider duplication. If a machine M (Fig. 17-1a) is nonredundant and is composed of three modules (M_1, M_2, M_3), it will have a certain reliability. Because of nonredundancy, a single failure in any of the modules causes the system to fail. If M is duplicated at the system level (Fig. 17-1b), by adding a duplicate M' consisting of modules M'_1, M'_2, M'_3 (for simplicity, ignore the problem of selecting the correct output), then no single failure in either M or M' can cause system failure. Neither can a multiple failure in either M or M' cause a system failure.

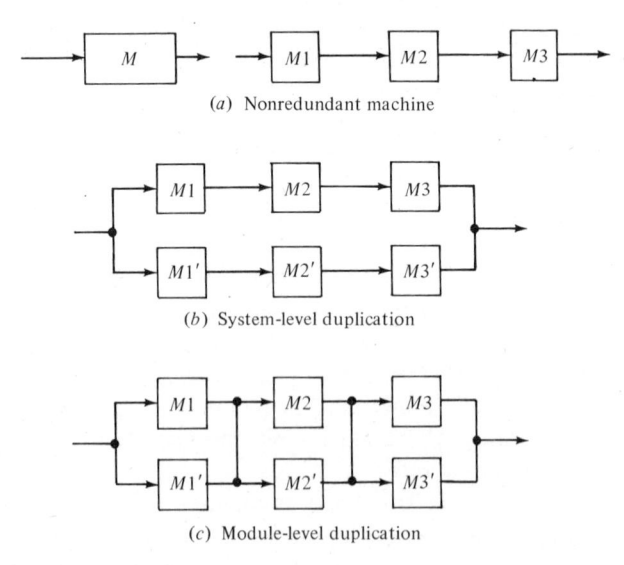

(a) Nonredundant machine

(b) System-level duplication

(c) Module-level duplication

FIGURE 17-1
Redundancy and effect of level of application of redundancy.

Any single failure in M plus any single failure in M', however, will cause system failure.

An alternative is to duplicate M at the module level (Fig. 17-1c). The same amount of hardware has been added. The system is invulnerable to all single failures and to all double failures, except simultaneous failures in any pair of duplicate modules. It is obvious that differences in level of application of redundancy have significant effects on the tolerance of a device to failure of multiple components.

One might believe that extrapolation of component-level redundancy would be appropriate. This is not necessarily the case. The preceding discussion ignored the problem of selecting the correct output, a function which requires additional circuitry. As the level of application is lowered, more circuitry is required, relative to the size of the modules, to make the selection. At some point, reliability of selection circuits begins to dominate system reliability.

Hardware redundancy in computers normally falls into two classes: spare switching and masking. Spare switching replaces a faulty element with an identical spare (a form of standby). Masking uses properly functioning neighbors of a faulty element to continue proper operation. Both spare switching and masking have been used at a variety of levels of application.

17-2-1 Spare Switching

Spare switching requires two operations: locating the faulty element and switching in the spare. One example of this technique is shown in Fig. 17-2. In this arrangement, the secondary element (spare) is used for fault detection by comparing it with the primary element. When the comparator indicates inequality between the two outputs, processing is halted and diagnostic programs are begun. If diagnostics indicate faulty operation, the primary unit has failed and the output switch is changed. If diagnostics indicate no fault, then failure is in the secondary circuit, the comparator, or a transient error in the primary, and a message is printed for the repairman. Spare switching gives significant increase in reliability with a minimal cost in additional circuits. This is a practical approach if maintenance is available.

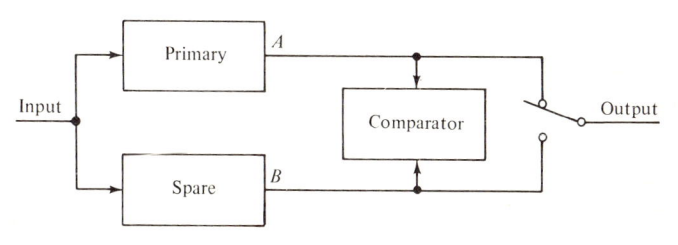

FIGURE 17-2
Spare switching technique of redundancy.

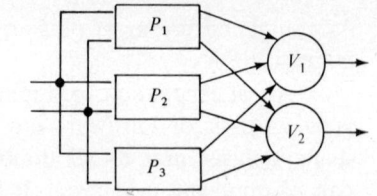

FIGURE 17-3
A voting circuit with three processors and two voters.

17-2-2 Voting Circuits

Voting circuits are a form of masking in which multiple replicates of a circuit each receive identical inputs. It is assumed that the output of a majority of these is the correct output. While there can be an arbitrary number of redundant circuits and voting circuits, the scheme will be illustrated (Fig. 17-3) with triplication of processors and duplication of voting circuits. A voter circuit functions as a device to sum the current and requires any two inputs to surpass the threshold. Output from the voter circuit is determined by "majority" vote of the three input signals. If processor P_2, for example, fails but P_1 and P_3 continue to function properly, two correct signals are sent to each voter which then gives the correct output response. If two processors fail, there is insufficient signal for the voter circuits to function.

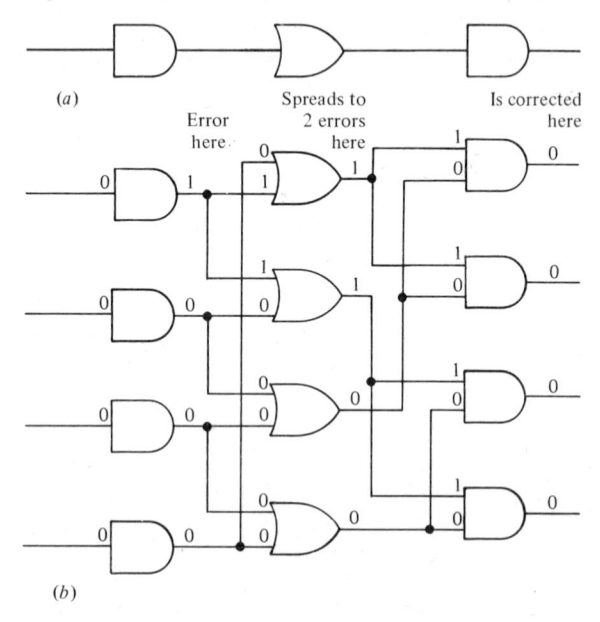

FIGURE 17-4
Schematic diagram for quadding.

17-2-3 Quadding

A second form of masking, known as "quadding," is based on three principles: (1) all circuits are in quadruplicate; (2) each error is corrected immediately downstream of the fault causing it; and (3) correction is made by good signals from neighbors of the faulty element. Figure 17-4 provides a simple example of such protection. If an error occurs either before or after an initial AND gate, that error is increased to two errors through the OR gate, and then is corrected (reduced to zero) by the correct signals at the final AND gates. This type of circuit is highly reliable and efficient.

If a comparator is used on the input and output sides of the error-correcting device, then the error can be detected as well as be corrected. Such information can be stored by the computer. If an abnormally high number of errors is detected for any given subsystem, this information can be printed out to the operator—giving location (and even possible cause) for maintenance action.

18

SAFETY FACTOR AND RELIABILITY

18-1 THE PROBLEM

The discussion in Sec. 8-1 deals with distributions of stress demand and available strength. Section 8-1 makes the point that if the distribution of stress demand and the distribution of available strength overlap (Fig. 8-2), failure will occur in the region of overlap. Consider the situation shown in Fig. 18-1 in which the stress-demand distributions all have the same mean but different standard deviations and the available-strength distributions have the same mean (different from the stress demand) but different standard deviations. Following the logic in Chap. 8, we would expect failure of very few devices in Fig. 18-1a, failure of more in Fig. 18-1b, and failure of a substantial percentage in Fig. 18-1c.

There has been recognition over the years (however implicit it may have been) that stress demands and available strengths do have variations. This has commonly been handled by using a safety factor or a worst-case approach. The results have not necessarily been bad. In fact, the result most often has been overdesign, i.e., conservative design. It is impossible, however, to readily determine, or even estimate, the amount of conservatism. It is equally impossible to predict those situations which may be marginal or underdesigned.

One can regard a "safety factor" as derating, in some sense. One differ-

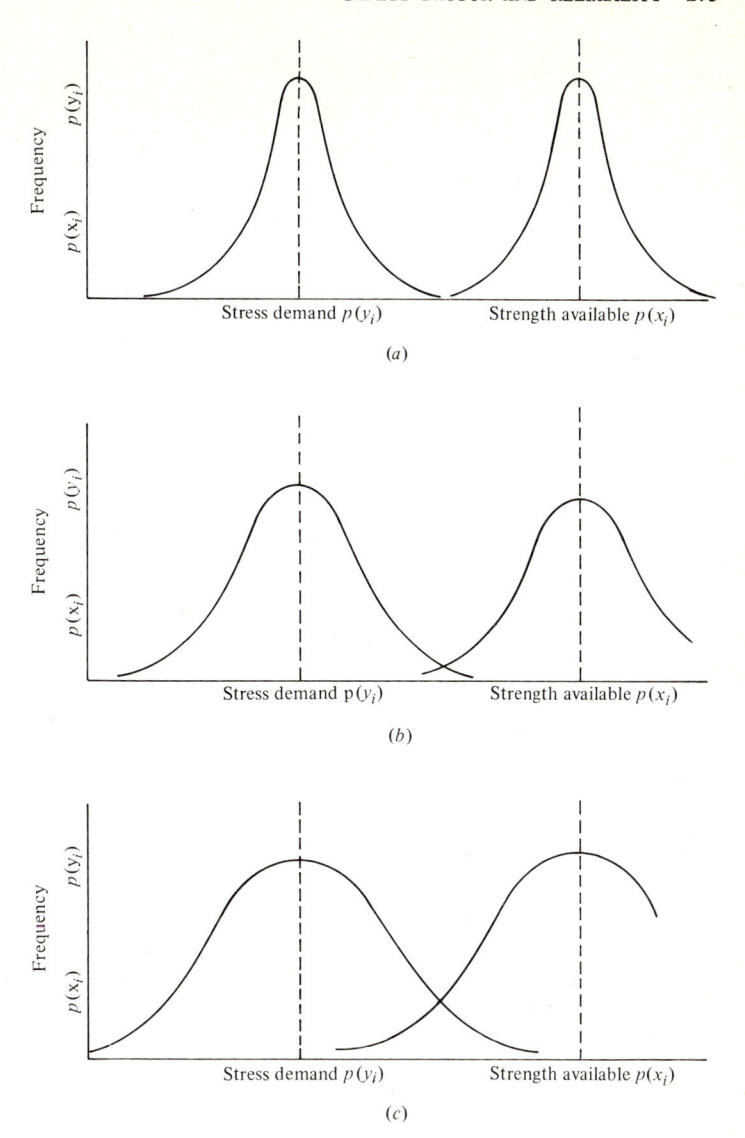

FIGURE 18-1
Distribution of stress demand and available strength (schematic) for three situations having the same values: (a) small standard deviations, (b) moderate standard deviations, (c) large standard deviations.

ence is that the amount of effective derating is unknown. Safety factor has been defined in a large variety of ways. One common definition is: ratio of ultimate (or yield strength) in a component to the allowable or working stress. In Fig. 18-1, all three cases would have the same safety factor, despite the obvious differences in the percent expected to fail.

Another definition of safety factor attempts to take the variations into account (in effect, one form of the worst-case approach). In this case, safety factor is defined as the ratio of mean strength S to the corresponding mean load L, i.e.

$$\text{Safety factor} \equiv \eta = \frac{S}{L} \quad (18\text{-}1)$$

Uncertainties cause variation in strength, ΔS, and in load, ΔL, from the mean values so that the lowest probable strength, $S - \Delta S$ and the highest probable load $L + \Delta L$ must satisfy the inequality $S \geq L$ for no failure, i.e.

$$(S - \Delta S) \geq (L + \Delta L) \quad \text{or} \quad S\left(1 - \frac{\Delta S}{S}\right) \geq L\left(1 + \frac{\Delta L}{L}\right)$$

Thus a minimum safety factor is

$$\eta = \frac{S}{L} = \frac{[1 + (\Delta L/L)]}{[1 - (\Delta S/S)]} \quad (18\text{-}2)$$

It is often believed that use of a safety factor greater than some preconceived magnitude, usually above 2.5, will result in no failure. Actually, with high safety factors, the failure probability may vary from a satisfactory low to an intolerable high. A safety factor of one implies, to many, that failure will occur 100 percent of the time because there is no safety factor. Actually, if strength and stress are normally distributed, failure will occur only 50 percent of the time.

It is well known that distributions exist in both the load (stress) requirement and the available strength. It is these distributions (as defined by mean values, standard deviations, and/or other parameters) with which the designer should be concerned. The safety-factor concept overlooks the facts of variability which may give different reliabilities for the same safety factor.

How would you relate the stress demand and the available strength, taking their variations into account so that you could predict the percentage which you expect to fail, on the average? In other words, how would you design to a desired reliability?

18-2 A SOLUTION

If the variables in a design problem are assumed to have normal distributions, then the algebra of normal functions (Appendix D) can be applied. Further, reliability is the probability that strength exceeds load, or $(S - L) \geq 0$. If S and L are normally distributed, then the difference between them is also normally distributed. This difference can be related to the standard-normalized variable, z, to give

$$z = \frac{S - L}{(S_S{}^2 + S_L{}^2)^{1/2}} \quad (18\text{-}3)$$

which is known as a "coupling equation" since it probabilistically relates the strength and load functions. In this context, z is called the coupling coefficient. Reliability (or probability of survival) is

$$R_e = \int_{-z}^{\infty} e^{-z^2/2}\, dz = \int_{-\infty}^{z} e^{-z^2/2}\, dz \qquad (18\text{-}4)$$

Once the value of z has been determined, the reliability can be determined directly from standard tables of normal functions. The probability of failure is $Q = 1 - R_e$.

18-2-1 Application to Mechanical Loading

All of the foregoing has been completely general. Let us apply the concept to the case of a hollow circular member which is subjected to both bending and torsion. Using the maximum-shear theory and the nomenclature given in Table 18-1, the single value (deterministic) solution for shear stress is

$$\tau = \frac{(M_b^2 + M_t^2)^{1/2}}{dR_o^3} \qquad (18\text{-}5)$$

In probabilistic notation this would be written

$$(\tau, S_\tau) = \frac{(M, S_M)}{(dR_o^3, 3dKR_o^3)} \qquad (18\text{-}6)$$

which is equivalent to

$$\tau = \frac{M}{dR_o^3} \qquad (18\text{-}7)$$

Table 18-1 NOMENCLATURE

	Symbols	
Name of variable	For variable (mean)	For standard deviation
Shear strength, psi	τ_0	S_{τ_0}
Bending moment, in.-lb	M_b	S_{M_b}
Twisting moment, in.-lb	M_t	S_{M_t}
Radius, in.	R	KR
Shear stress, psi	τ	S_τ
Combined moment, psi	M	S_M
Unit twist, rad/in.	θ	S_θ
Deflection, in.	δ	S_δ
Radius ratio	$R_o/R_i = C$	
Polar moment of inertia	I_P	

$I_P = \pi/2(R_o^4 - R_i^4) = dR_o^4$

where $d = \pi/2 \left(\dfrac{c^4 - 1}{c^4} \right)$

and

$$S_\tau = \frac{1}{dR_o^3}\left(\frac{9K^2M^2 + S_M^2}{1 + 9K^2}\right)^{1/2} \quad (18\text{-}8)$$

where

$$M = [(M_b^2 + M_t^2)^2 + 2(M_b^2 S_{M_t}^2 + M_t^2 S_{M_b}^2 + S_{M_b}^2 S_{M_t}^2)]^{1/4} \quad (18\text{-}9)$$

$$S_M = [(M_b^2 + S_{M_b}^2 + M_t^2 + S_{M_t}^2) - M^2]^{1/2} \quad (18\text{-}10)$$

Equation (18-3), the coupling equation, then becomes

$$z = \frac{\tau_0 - \tau}{(S_{\tau_0}^2 + S_\tau^2)^{1/2}} \quad (18\text{-}11)$$

Solution of this equation for the radius gives

$$R_o^6 - \frac{2\tau_0 M}{d(\tau_0^2 - z^2 S_{\tau_0}^2)} \cdot R_o^3 + \frac{M^2[1 - 9K^2(z^2 - 1)] - z^2 S_M^2}{d^2(1 + 9K^2)(\tau_0^2 - z^2 S_{\tau_0}^2)} = 0 \quad (18\text{-}12)$$

EXAMPLE 18-1   As an application of Eq. (18-12), consider a combined bending and torsion problem in which the parameters have the values given in Table 18-2. It is obvious that solution of Eq. (18-12) is much more tedious than solution of Eq. (18-5).   Equation (18-12) can, however be programmed readily to yield a computer solution.   The solution has been made with the results shown in Fig. 18-2.   The straight-line relationships (on probability paper) substantiate at least two results which are intuitively expected: (1) Since all the distributions involved are assumed to be normal distributions, the straight line would be expected.   (2) As higher degrees of reliability are required, the shaft size must increase greatly.

The effects of various standard deviations on the required radius are also of interest.   In this problem, the required radius is a function of shear strength, bending moment, and twisting moment.   Table 18-3 shows the effect on the required radius for two different reliabilities, with each of the four parameters having its standard deviation varied while the other three are held constant.   Figure 18-3

Table 18-2 PARAMETERS FOR EXAMPLE 18-1

Variable	Mean	Range	Standard deviation
Shear strength, (τ_0) psi	45,000	±6600	2200
Bending moment, M_b, in.-lb	60,000	±9000	3000
Twisting moment, M_t, in.-lb	40,000	±6000	2000
Radius, R, in. ($k = 0.015$)	R	±0.045R	0.015R

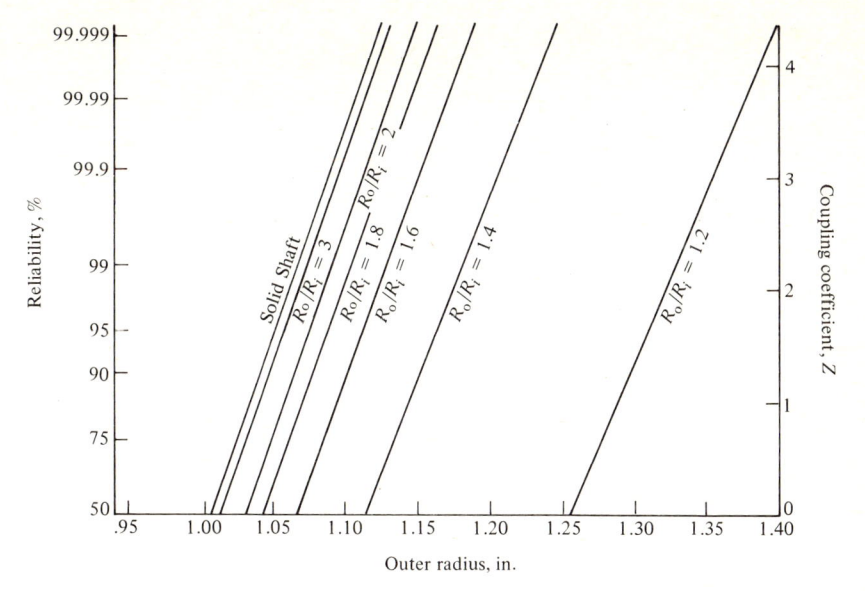

FIGURE 18-2
Relationship between radius and reliability for various radius ratios.

shows this effect for two reliabilities as a function of variation in standard deviation of the shear strength.

As expected, the required radius increases as the standard deviation of any one of the four parameters increases. This increase is clearly not linear (Fig. 18-3). It should be noted that the increase in radius is necessary to maintain a given reliability. The safety factor, based on mean values, would be the same in each case in Table 18-3. Thus, in effect, for a given safety factor the reliability decreases (or

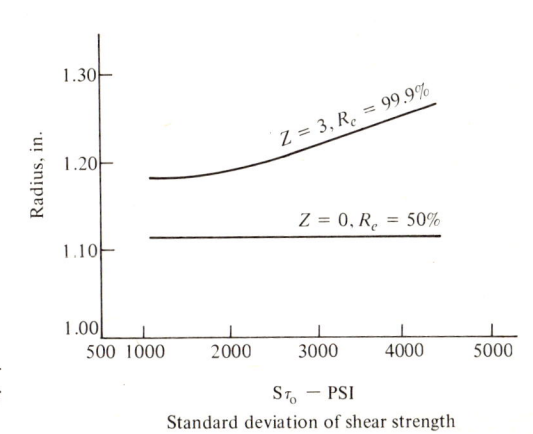

FIGURE 18-3
The effect on outside radius of varying the standard deviation of the shear strength for $R_o/R_i = 1.4$.

Table 18-3 EFFECT ON RADIUS OF VARIATIONS IN STANDARD DEVIATION[a]

Variable	Reliability %	Standard deviation Numerical	Standard deviation %	Mean outside radius, in.
Shear strength, τ_o 45,000 psi	50			1.1133
	99.9	1100	2.5	1.1820
		2200	4.9	1.1982
		3300	7.3	1.2287
		4400	9.8	1.2642
Bending moment, M_b 60,000 in.-lb	50			1.1133
	99.9	1500	2.5	1.1948
		3000	5.0	1.1982
		4500	7.5	1.2098
		6000	10.0	1.2207
		7500	12.5	1.2329
Twisting moment, M_t 40,000 in.-lb	50			1.1133
	99.9	1000	2.5	1.1969
		2000	5.0	1.1982
		3000	7.5	1.2028
		4000	10.0	1.2054
		5000	12.5	1.2087
Radius, R_o, in.	50			1.1133
	99.9	$0.0025\,R_o$	0.25	1.1871
		$0.005\,R_o$	0.5	1.1884
		$0.010\,R_o$	1.0	1.1933
		$0.015\,R_o$	1.5	1.1982
		$0.020\,R_o$	2.0	1.2100

[a] For parameters listed in Table 18-2 with $R_o/R_i = 1.4$.

the failure probability increases) as the range ($\pm$ three standard deviations) of any one or all parameters increases. ////

EXAMPLE 18-2 For additional insight into the situation, consider the problem of a dc motor which produces 75 hp at 1150 rpm. The rotor weighs 400 lb with its solid shaft and is supported on bearings which have centers 30 in. apart. Power is taken off through a flexible coupling. Determine the shaft size required, the maximum linear deflection, and the unit twist using AISI 1040 steel. The parameters are stated in Table 18-4.

This problem was solved in five different ways with the results given in Table 18-5. In each case, the values for unit twist and maximum deflection were found

Table 18-4 PARAMETERS FOR EXAMPLE 18-2

Variable	Mean value	Range, %	Standard deviation, %
Horsepower	75	±12	4
Rev. per min	1150	±12	4
Weight, lb	400	±15	5
Length, in.	30 plus	± 6	2
Radius, in.	R	± 6	2
Shear strength, psi	27,000	±24	8
Modulus of elasticity E, psi	30×10^6	± 6	2
Modulus of rigidity G, psi	12×10^6	± 6	2

by using the size radius determined. The first set of results was determined by using mean values, in effect, a reliability of 50 percent. In the second case, a minimum safety factor of 1.5 was determined from Eq. (18-2). The third set of results was determined on the basis of the worst possible combination of the parameters. The fourth and fifth sets of results were determined on the bases of 99.9 percent and 99.997 percent reliability, respectively.

From Table 18-5, it is obvious that the smallest radius calculated is that determined from the mean values. It is also true that this has one chance in two of failing. It is equally obvious that the largest radius calculated is that determined from the worst-case treatment.

Comparison of results from the three single-value methods with those from the two degrees of reliabilities makes it obvious that both the minimum safety factor and

Table 18-5 SOLUTIONS FOR EXAMPLE 18-2

	Radius, in. R	Stand. dev. KR	Range	Unit twist, rad/in. θ	Stand. dev. S_θ	Range	Maximum deflection, in. δ	Stand. dev. S_δ	Range
Mean values, Reliability = 50%	0.469	—	—	0.00453	—	—	0.123	—	—
Minimum safety factor of 1.5	0.547	—	—	0.00258	—	—	0.0704	—	—
Worst case	0.558	—	—	0.00306	—	—	0.0900	—	—
Reliability = 99.9%	0.527	0.010	±0.031	0.00283	0.000233	±0.000699	0.0773	0.00531	±0.0159
Reliability = 99.997%	0.550	0.011	±0.033	0.00238	0.000202	±0.000607	0.0653	0.00424	±0.0127

worst-case methods give sizes which are close to the sizes determined from the probabilistic method. The major point, however, is that the degree of reliability (the amount of under- or overdesign) is not known. Overdesign is not necessarily wrong, but it is highly desirable to know the degree of overdesign and thus be able to estimate the price to be paid in money, weight penalty, etc. It is interesting to note that an actual motor of essentially this size had a shaft radius of 1.406 in. ////

18-2-2 Some Comments

Comparison of the calculated radii in Table 18-5 with the measured size of an actual motor makes it appear that the actual shaft is vastly over designed. No doubt it is somewhat overdesigned, but there are other factors which have been neglected. These are such things as stress-concentration factors, load factors (dynamic, impact, etc.), temperature factors, forming- or manufacturing-stress factors (residual stresses, surface treatment, heat treatment, assembly), corrosion-stress factors, and notch-sensitivity factors. All of these factors influence the required size. They were not included so that the central idea of the probabilistic approach might be clearer. The probabilistic approach certainly does not eliminate these factors, for they must be included. Stress-concentration factors are normally used to increase the effective magnitude of applied loading. Thus, appropriate stress-concentration factors should be applied to the given bending and twisting moments to determine the effective values to be substituted into the equations. Temperature factors should be applied to the strength of the material to be used. In like manner, all other necessary modification can be integrated into the solution.

Some of the references present solutions to other configurations of mechanical loading. Reference 18-1 applies the basic concept covered in this chapter to a variety of other situations.

19
RELIABILITY ALLOCATION

19-1 THE PROBLEM

The concept of an unmanned automated geophysical observatory emerged several years ago as a plausible alternative to manned stations for Antarctic research. The principal motivating factor for this concept was the continuing burden of high cost and great effort associated with life support in manned stations. The fact that automated laboratories, similar in function and complexity to their Antarctic counterparts, were already successfully completing missions in space made the concept appear feasible.

The purpose of the station is to make a wide range of scientific measurements in Antarctica. Selection of experiments and experimenters is at the discretion of the National Science Foundation. To assess the requirements for the various subsystems, however, it is necessary to specify a typical list of experiments. This list of experiments was chosen with two criteria in mind. First, there must be adequate scientific interest in the experiment to ensure that the data will be used. Second, the experiment must be one which can be operated without the presence of a man (e.g., experiments which require continual adjustments must be improved or eliminated).

19-1-1 Requirements

Stated briefly, the ideal automated Antarctic observatory would operate unattended for one year (it would be serviced annually during the austral summer), would transmit data at moderately high rates to the United States with little delay, would be easily transported by air and quickly assembled in the field, and would have flexible facilities (such as data processing, timing signals, environmental control, and power supplies) which would permit periodic alteration of the experiments. The observatory would collect most types of geophysical data presently being obtained at manned Antarctic stations.

The experiments studied and their requirements are summarized in Table 19-1. The combined requirements are data rates of 600 bits/s plus an analog bandwidth of 2.5 kHz, volume of 15 ft³, and power consumption of 65 W. These experiments must be protected from the hostile Antarctic environment, with its cold temperatures (0 to –88°C), high winds (up to 150 mph), and snow accumulation (up to 5 ft/year in some areas).

The need for a continuous high data rate will require the use of a relay

Table 19-1 SUMMARY OF STATION REQUIREMENTS

Item	Bit rate, bits/s	Power, W	Volume, in.³	Comments
1 Telemetry		37.5	4,100	
(a) Dish, rec. trans.	–	–	6 ft diam.	
(b) Receiver	–	1.0	200	
(c) Transmitter	–	15	1,700	
(d) Sampler	–	3	200	
(e) A/D Converter	–	1.0	200	10 bits/word
(f) Clock	–	9	1,200	1 part in 10^{10}
(g) Command decoder	–	1.5	200	
(h) dc/dc Converter and regulator	–	7.0	400	
2 Experiments	400	65	7,900	
(a) Riometer	10	10	600	30 × 30 ft ground plane
(b) Aurora and anglow	5	5	0	6-in. diam. dome
(c) Cosmic rays	5	5	1,700	
(d) Geomagnetism	1	10	800	
(e) Micropulsations	300	6	600	3, 6-ft loops
(f) Meteorology	0.1	5	200	
(g) Seismology	80	5	800	
(h) VLF-SPECTRUM	2.5 kHz	5	1,700	Analog bandwidth
(i) VLF-Phase	2	10	800	
(j) Housekeeping	1	5	500	
3 Beacon (radio + strobe)	–	12	200	Beacon strobe and antenna outside
4 Totals	400 + 2.5 kHz	102.5	12,000	Volume = 15 ft³ including packaging allowance

through a synchronous satellite, either NASA's ATS satellites or the commercial Intelsat III satellites (costing between $7,000 and $15,000/year for the desired data rates).

19-1-2 Configuration and Installation

The system which best matches the Antarctic environment consists of an insulated instrument capsule elevated above the snow, a field support van which serves as the shipping container for the entire observatory as well as the field-site servicing and calibration facility, and thermoelectric radioisotope generators and/or wind-powered generators.

The instrument capsule should be mounted 15 ft above snow level in order to reduce the rate of snowdrift accumulation around the observatory. The van should be provided with runners so that it can be located directly under the instrument capsule during installation and annual servicing or moved to a remote location for system checkout. All circuitry is mounted in a 19-in. instrument rack which can be raised or lowered into the instrument capsule.

The interior of the instrument capsule should be maintained at $0 \pm 10°C$ in order to eliminate mechanical fatigue effects on circuit components due to temperature fluctuations. Temperature will be controlled by mechanically altering the thermal conductance through the capsule walls with a bimetal switch. The interior will be tightly sealed from the wind and windborne snow.

Mounted on the instrument capsule is a 6-ft parabolic antenna for communicating with a geostationary satellite. The antenna can be moved up to 10 degrees in any direction upon command relayed from the United States via satellite. This movement will be used to accommodate small, long-term shifts in the position of the satellite or of the observatory, which is mounted on snow pack.

The entire station should be designed to be transported to its location and installed with three flights of an LC-130 transport from McMurdo Sound. The first two flights will bring in auxiliary equipment such as a tractor, diesel generator, fuel, temporary shelters, food, and other support material sufficient for a two-week period. The third flight will bring in all support personnel, the van and station, and the personnel directly associated with the station installation and checkout. After the two-week installation and checkout period, two additional flights would be necessary to return the personnel and equipment to McMurdo sound.

19-1-3 Data Transmission

Many approaches are possible for the retrieval of scientific data. The great volume of data eliminates the possibility of recording it on site and retrieving it during the annual maintenance visit. Microwave relay to a manned Antarctic station would be far too costly (each relay station would itself be as expensive as the scientific station). High-frequency transmission to a manned station would require very high power and would be interrupted by the polar blackout that occurs during moderate and heavy auroral activity.

The best solution is use of satellites to relay data to the United States. The few polar orbiting satellites available do not have sufficient data-storage capacity and do not have scheduled programs to provide guaranteed service during the planned operation of the Antarctic station. Synchronous satellites on the other hand, transmit data directly to the United States and thus need no storage.

It is thus recommended that the station use a 6-ft parabolic antenna and 4-W TWT amplifier to transmit the 600-bits/s plus 2.5-kHz data through an ATS satellite for the first year or two of operation. Following that (since ATS, as an experimental satellite, cannot provide continued service) the data would be transmitted through a commercial Intelsat III satellite.

In addition, a low data rate, 10 bits/s or less, would be transmitted from the United States through the satellite to the Antarctic to provide control of the experiments and station subsystems. This command link can be used to change the data rate of the whole station or of each individual experiment. Thus the station data rate may be kept at a maximum consistent with a specified error probability. Also the experiment-data rates can be modified to study selected events in more detail.

Scientific data are relayed through a telemetry link to the United States. At the station in the United States the incoming scientific data are monitored for quality. Housekeeping data are monitored to ensure that all subsystems are operating properly. The data are also recorded on a tape recorder. This master tape is then decommutated on an off-line computer and tapes are made for each experimenter. These tapes include the basic data, timing information, and pertinent housekeeping data (sensor temperature, power, etc.).

19-1-4 Experiment Interface

The station must be designed to minimize the problem of mating the various experiments to it. The experiments must be integrated into the design thermally, mechanically, and electrically. The station must provide standard power-supply voltages and data interfaces through standard connectors and an environment which is carefully controlled in both temperature and humidity. This standardization will result in more reliable interfaces and shorter experiment integration times.

Since on-site computations are not required in the experiments and since the initial experiments can be carried out using elementary open-loop control, a computer is not required. The computer functions are performed by special-purpose hardware, i.e., a data encoder, command decoder, and sequencer.

The hardware required for the system has been divided into five major parts:

1 Analog switches and A/D converters
2 Buffer memory

3 Sample memory
4 Output-control logic
5 Switch-driving circuitry

The required number of channels is 27 and the highest sample rate is 10 samples per second. It was believed that 64 channels and 100 samples per second would allow sufficient capacity for future expansion, and these values are adopted for design purposes. The input to the A/D converter will thus be 6400 samples per second, allowing 156 μs per conversion. This can be accomplished with a fairly simple off-the-shelf A/D converter. The analog sampling circuitry is standard and may be purchased on printed-circuit cards ready for use. The rest of the circuitry is unique and must be designed in detail. A summary of the necessary hardware is given in Table 19-2.

19-1-5 Reliability Estimate

Reliability is an important consideration for the unmanned station since flight operations are possible during only three months of the year. If the station fails during the other nine months it cannot be repaired until the following summer. Thus a station failure will result in loss of data, loss of face, and loss of experimenter enthusiasm. Still, the situation is much better than that in space where the capital investment in the spacecraft is also lost upon failure.

Calculate the reliability of the data handling and transmission system for an operating period of one year. Do you consider this an acceptable level of reliability? If not, what level of reliability would you accept? What assumptions underlie your decision as to acceptable level? How would you improve the reliability to a level which you will accept?

Table 19-2 HARDWARE SUMMARY

Subsystem	Power, W	Volume, in.^{3}	Reliability, failures/10^6 h	Cost, dollars
1. Data encoder	4.0	200	42.7	21,000
2. Command decoder	1.5	175	16.4	15,000
3. Antenna	—	External	12.5	5,000
4. Diplexer	—	50	10.0	2,000
5. Transmitter	7.5	1200	85.0	12,000
6. Modulator	0.5	200	10.0	3,000
7. Command receiver	1.0	200	15.0	12,000
8. Frequency standard	9.0	1200	10.0	1,500
9. dc/dc converter	4.0	400	10.0	3,000
10. dc regulator				
Totals (non-VLF)	30.5	3625	211.6	$74,500
(with VLF)	36.5	3625	211.6	$74,500

19-2 ONE SOLUTION

The first step in a reliability analysis is to assess the reliability of each subsystem. In this situation, the data have already been given, in Table 19-2. Taking the simplest possible configuration, i.e., one of each subsystem in series, we find the effective total failure rate is $211.6/10^6$ h for an MTBF of 4740 h. Thus on the average, we can expect the system to function for 4740 h (8760 h/year). This gives a reliability of 0.155 which means there is about one chance in six of the station operating for one year. In terms of the objectives and constraints given in Sec. 19-1, this is much too low a reliability.

There are five methods for improving reliability:

1 Component part improvement
2 Stress level reduction
3 Circuit simplification
4 Improvement of manufacturing techniques
5 Redundancy

All these methods should be included in a full reliability program, but redundancy is the only effective method available in the systems-planning stage.

A computer program (19-1) was written which found the optimum for each subsystem given the MTBF, cost of each subsystem, and dollar value of the data for a year. This approach assumed that the system will be used many times, so that the average lifetime will approach the expected average lifetime. A look at the subsystem-failure rates given in Table 19-2 shows that the transmitter is a good candidate for improvement. The principal source of transmitter failures is the output traveling-wave tube (TWT). It was decided to apply parallel standby redundancy to this TWT with switchover done on command. With this modification, the subsystem-failure rates were put into the program. The subsystem costs used in the program are the incremental costs of producing a second unit given the existence of the first. For subsystems which require development, a second unit will cost between one-third and two-thirds the cost of the initial unit. The results of this computer program are shown in Fig. 19-1. C_1 is the value of a year's data; C_2 is the cost of the nonredundant system. In using Fig. 19-1, the value of C_2 used should be the average annual cost where the average is taken over the number of stations built and over the useful life of the station. Table 19-3 illustrates the use of the graph. It was assumed that only one station was built, the electronic subsystems (C_0) cost \$100,000, the data for one year were worth 5 million dollars, and maintenance costs were negligible (C_T is total cost).

From Table 19-3, it is obvious that amortization over a longer period of time decreases annual cost and increases reliability. For useful lives up to 3 years, the optimum reliability is 0.513; for useful lives between 6 and 20 years, the optimum reliability is 0.922. Choice of an appropriate level of redundancy requires specification of both C_1 and useful life of the system. For this study (Ref. 19-1) a useful life of over 6 years was assumed and the value of C_1 was

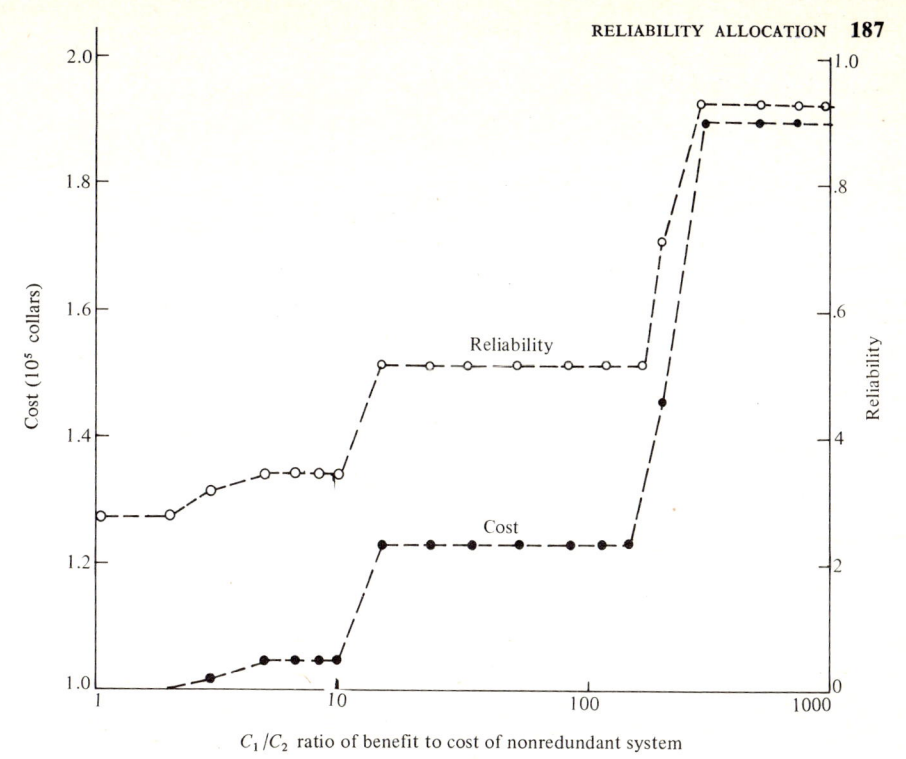

FIGURE 19-1
Results of computed cost and reliability analysis.

assigned at 5×10^6, leading to a reliability of 0.922. To obtain this level of reliability, all subsystems were duplicated and some were triplicated. The total cost of this level of redundancy was expected to be $110,000.

Obviously, Jenny and Lapson (19-1) considered a reliability of 0.922 as acceptable. One can agree or disagree, as this is clearly a matter of value judgment. This level of reliability may not be unreasonable upon realizing that the probability of operating 11 out of 12 months is 0.916.

Table 19-3 COST AND RELIABILITY ANALYSIS

Useful life (years)	C_1/C_2	R	C_T/C_0
1	50	0.513	$123,000
3	150	0.513	123,000
4	200	0.712	146,000
6	300	0.922	190,000
10	500	0.922	190,000

[It may be of interest that the system was built as simply as possible with components which had been qualified in the space program. There was extended burn-in to be certain the system was in useful life. The observatory and related telemetry functioned unattended for the first year with no known failures. At the beginning of the second year, some changes were made to provide better switching. During the second year of operation, power was supplied from a central source which experienced a surge which, in turn, "blew" part of the system. Although the system was properly designed to function in a remote location with its own power source, it apparently was not properly fused for self-protection from a large power surge (19-2).]

19-3 REDUNDANCY ALLOCATION

A more general approach to an exact solution to the problem of redundancy allocation in a multistage system has been developed by Kettelle (19-3). In this situation, an M-stage system is considered. This system is operational if, and only if, each stage (stages effectively in series) has at least one operational component out of one or more components in parallel. The cost of a component in stage i is c_i and has availability $a_i (i = 1, 2, \ldots, M)$. Availability is the probability that a component or system is operational at any given point in time. Failure behavior, repair, and maintenance are assumed independent among all components and subsystems. Thus system availability is

$$A = \prod_{i=1}^{i=M} [1 - (1 - a_i)^{n_i}] \qquad (19\text{-}1)$$

where n_i is the number of parallel components in stage i. The cost of the system is

$$C = \sum_{i=1}^{i=M} c_i n_i \qquad (19\text{-}2)$$

If the system availability requirement is R, the problem then is to determine a least-cost configuration, i.e. $n_1, n_2, \ldots n_M$, that gives $A \geqq R$.

The algorithm developed by Kettelle constructs dominating sequences for successively larger groups of stages until the entire system is included. For a given group of redundancy configurations, defined over a given set of stages, an availability-cost sequence $[(A_0, C_0), (A_1, C_1) \ldots]$ is defined to be a dominating sequence if (A_j, C_j) for $j = 1, 2, \ldots$ is a least-cost entry in the group with reliability exceeding A_{j-1}. Likewise, one configuration dominates another if it has (a) more availability and no more cost, or (b) no less availability with less cost. A dominating sequence contains only configurations that are undominated. The given reliability-cost sequence for each individual stage is (trivially) dominating for that stage.

The steps of the procedure are:

1 Plan successive pairings of stages or groups of stages, each new combination consisting of two previous stages or combinations. Continue until the entire system has been combined. There will be alternative possibilities of pairings. It is not clear how much is gained by ingenious pairings, but there are simplifications if both members of the pair have equal steps in reliability or cost. In any event, there will be $M - 1$ pairings.

2 For each stage in the system, determine the minimum number of components required to attain the availability R at that stage alone. These minimum requirements for each stage are known as "base" requirements. The rest of the computation focuses on additional requirements and additional costs above the base. (If there is a possibility of accepting a lower value of R than originally specified, this should be included at this point, as later inclusion may require reworking the entire problem.)

3 Prepare a table for the first pair of stages. Across the top, post the costs c_{1j} and unavailabilities b_{1j} ($b_{ij} = 1 - a_{ij}$) corresponding to successive additions of components above the base requirements (labeled O) for stage 1, i.e., the first b_{1j} posted should be less than $Q = 1 - R$. Post successive steps along the top until the stage unavailability promises to be low enough to support the system requirement. The table can be extended later, if necessary. Post comparable information for stage 2 down the left-hand side of the table.

4 Construct the dominating sequence for the first two stages (first pair) beginning with the entry in the upper left corner. This will be the first term (its unavailability may not necessarily be less than the system unavailability Q, since it is clearly the least-cost entry. Compute and post costs (c_{ij}) and unavailabilities (b_{ij}) of other entries only as needed.

If element E_{ij}, at the intersection of row i and column j, is an element of the dominating sequence, then the next element will be restricted to rows $r \geq i$ and columns $s \geq j$. From each such row or column, the candidate for the next member of the dominating sequence is the least-cost element (in that row or column) with less unavailability. The next element in the sequence selected from all candidates is the least cost. In some cases, the rest of a row or column can be eliminated. For example, if the unavailability B_{ij} is less than the unavailability b_{2k} posted at the left of a previous row k, then all entries in row k which cost more than C_{ij} can be rejected.

5 Starting with 0, number the entries (in order of cost) that cannot be dominated and whose unavailability is less than the system unavailability Q. Continue until the dominating sequence has progressed to an availability which supports the system requirement.

6 Proceed to the next level of combinations, using availability and cost functions given by the previous dominating sequence as inputs. Continue through successive levels until all stages have been combined. The resulting dominating sequence is the progressively more expensive (but progressively more reliable) configuration.

This allocation algorithm is shown in flow-chart form in Fig. 19-2.

190 INTRODUCTION TO RELIABILITY IN DESIGN

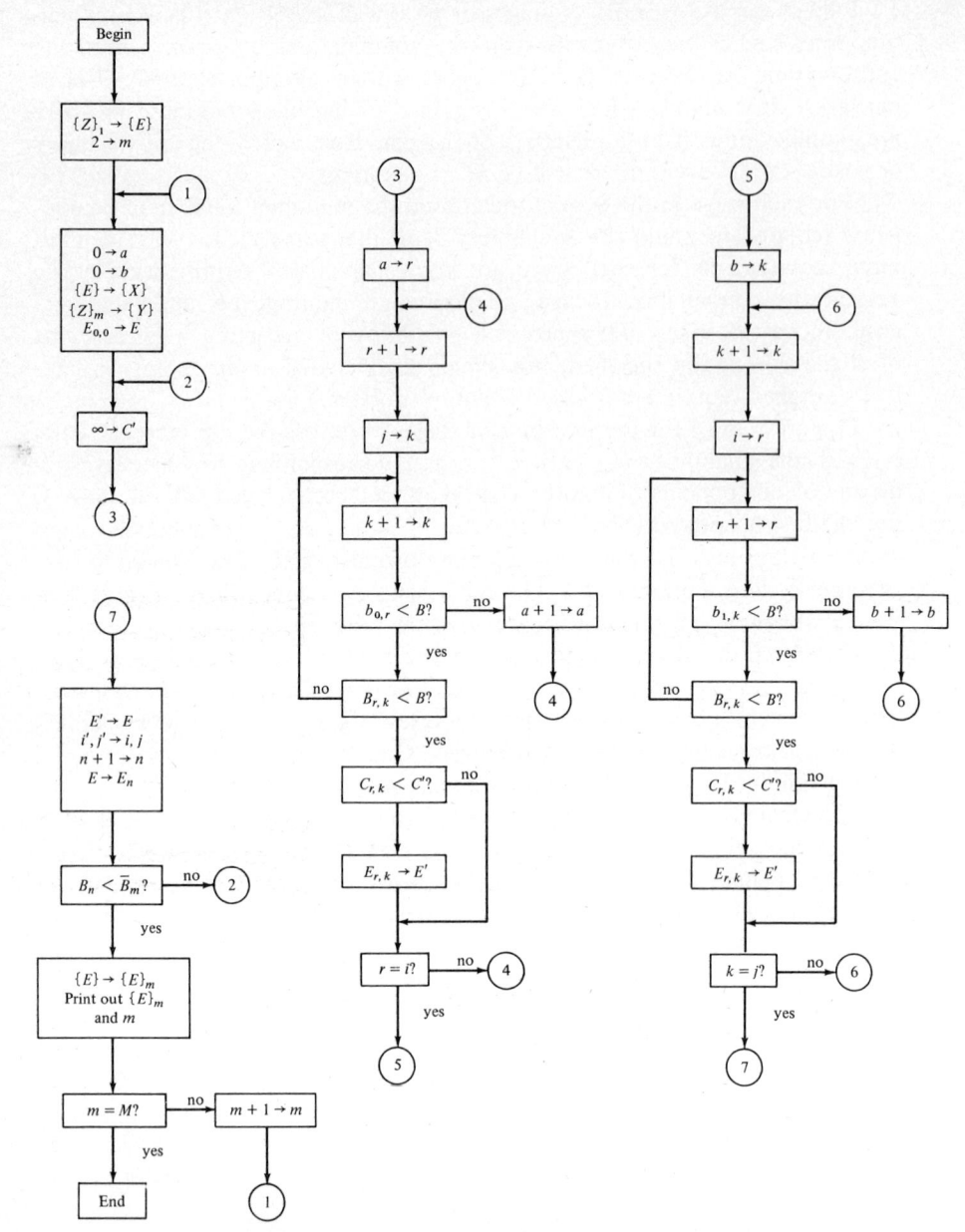

FIGURE 19-2
Computer flow chart for allocation algorithm.

Table 19-4 COMPONENT CHARACTERISTICS

Stage	1	2	3	4
Component cost	1.2	2.3	3.4	4.5
Component availability	0.8	0.7	0.75	0.85

EXAMPLE 19-1 Consider a four-stage system having the component characteristics indicated in Table 19-4. What is the appropriate configuration to give a system availability of 0.99?

From step 1, the following pairings are possible. (The right-hand pairing will be used here.)

$$\left.\begin{matrix}1\\2\end{matrix}\right\}\left.\begin{matrix}\\3\\4\end{matrix}\right\} \qquad \left.\begin{matrix}1\\2\end{matrix}\right\}\left.\begin{matrix}\\3\end{matrix}\right\}\\4$$

Step 2 calculates the base requirements giving the results shown in Table 19-5.

Table 19-6 was developed by step 3, combining stages 1 and 2. Stage 1 has a component cost of 1.2, a component unavailability of 0.2, and a base requirement of 3 components. Thus

$$C_{1j} = (3+j)1.2$$
$$b_{1j} = (0.2)^{3+j} \qquad (j = 0,1,2, \ldots)$$

Similar calculations are made for stage 2.

Step 4 constructs the dominating sequence. For this example, the cost of the combination (C_{ij}) and unavailability (B_{ij}) achieved by the combination at row i and column j are:

$$C_{ij} = C_{2i} + C_{1j}$$
$$B_{ij} = b_{2i} + b_{1j} - b_{2i} \cdot b_{1j}$$

(If both b_{2i} and b_{1j} are small, the product may be negligible.) These summations are entered in the appropriate place in Table 19-6. The boundary of the rejected (dominated) combinations has been indicated by shading in Table 19-6.

Table 19-5 BASE REQUIREMENTS

Stage	1	2	3	4
Number of components required	3	4	4	3
Basic cost	3.6	9.2	13.6	13.5
Stage unavailability	0.0080	0.0081	0.0039	0.0034

Table 19-6 COMBINATIONS OF STAGES 1 AND 2

Steps for stage 1

	i c b	0 3.6 .0080	1 4.8 .0016	2 6.0 .00032	3 7.2 .000064	4 8.4 .000013	5 9.6 .000003	6 10.8
Steps for stage 2	0 9.2 .0081	12.8 .0161	(0) 14.0 .0097	(1) 15.2 .0084	16.4			
	1 11.5 .00243	15.1	(2) 16.3 .0040	(3) 17.5 .0027	18.7			
	2 13.8 .000729	17.4	(4) 18.6 .0023	(5) 19.8 .00105	(6) 21.0 .00079	22.2		
	3 16.1 .00022		20.9	(7) 22.1 .00054	(8) 23.3 .00028	(9) 24.5 .00023	25.7	
	4 18.4 .000066			24.4 .000386	(10) 25.6 .00013	(11) 26.8 .000079	(12) 28.0 .000069	29.2
	5 20.7 .000020				27.9 .000084	(13) 29.1 .000033	30.3	31.5
	6				30.2	31.3		

i = number of step
c = cost above basic cost
b = unavailability
( ) = dominating sequence (in order)

RELIABILITY ALLOCATION 193

Table 19-7 COMBINATIONS OF STAGES 3 AND 4

Steps for stage 3

i	0	1	2	3	4	5
c	13.6	17.0	20.4	23.8	27.2	30.6
b	.0039	.00097	.00024	.000061	.000015	.000004

Steps for stage 4

0 13.5 .0034	(0) 27.1 .0073	(1) 30.5 .0044	(2) 33.9 .0036	37.3		
1 18.0 .00051	31.6 .0044	(3) 35.0 .00148	(4) 38.4 .00075	(5) 41.8 .00057	45.2	
2 22.5 .000076	36.1	39.5	(6) 42.9 .00032	(7) 46.3 .000137	(8) 49.7 .000091	53.1
3 27.0 .000011			47.4	(9) 50.8 .000072	(10) 54.2 .000026	57.6
4 31.5 .0000017				55.3	58.7	62.1

i = number of step
c = cost above basic cost
b = unavailability
( ) = dominating sequence (in order)

194 INTRODUCTION TO RELIABILITY IN DESIGN

Step 5, shown by lines in Table 19-6, lists the undominated entries in order of cost. In this example, this has been carried to 13 terms, which may be more than necessary.

The first phase of step 6 combines stages 3 and 4, to give Table 19-7, in the same manner as stages 1 and 2 were combined to give Table 19-6. In this problem, the last step is to combine the results of the previous pairings to give Table 19-8. It might be noted that increases in cost and availability are irregular in the final pairing. This does not affect the effective working of the procedure. A dominating sequence often contains a relatively inefficient step (a small change

Table 19-8 COMBINATIONS OF LAST PAIRING TO GIVE TOTAL SYSTEM

Steps for stages 1, 2

	i	0	1	2	3	4	5	6	7	8	9
	c	14.0	15.2	16.3	17.5	18.6	19.8	21.0	22.1	23.3	24.5
	b	.0097	.0084	.0040	.0027	.0023	.00105	.00079	.00054	.00028	.00023
Steps for stages 3, 4	0 27.1 .0073	41.1	42.3	43.4	(0) 44.6 .0100	(1) 45.7 .0096	46.9 .0084	48.1 .0081	49.2	50.4	51.6
	1 30.5 .0044	44.5	45.7 .0128	(2) 46.8 .0084	(3) 48.0 .0071	(4) 49.1 .0067	(5) 50.3 .0055	(6) 51.5 .0052	52.6		
	2 33.9 .0036	47.9	49.1	50.2 .0076	51.4 .0063	52.5 .0059	53.7 .0047	54.9			
	3 35.0 .00148	49.0	50.2	51.3 .0055	(7) 52.5 .0042	(8) 53.6 .0038	(9) 54.8 .0025	(10) 56.0 .0023	(11) 57.1 .0020	(13) 58.3 .00176	59.5
	4 38.4 .00075	52.4	53.6	54.7	55.9 .0035	57.0	(12) 58.2 .0018	(14) 59.4 .0015	(15) 60.5 .0013	61.7	62.9
	5 41.8 .00057	55.8		58.1	59.3	60.4	61.6	62.8	63.9	65.1	66.3

i = number of step
c = cost above basic cost
b = unavailability
( ) = dominating sequence (in order)

RELIABILITY ALLOCATION   195

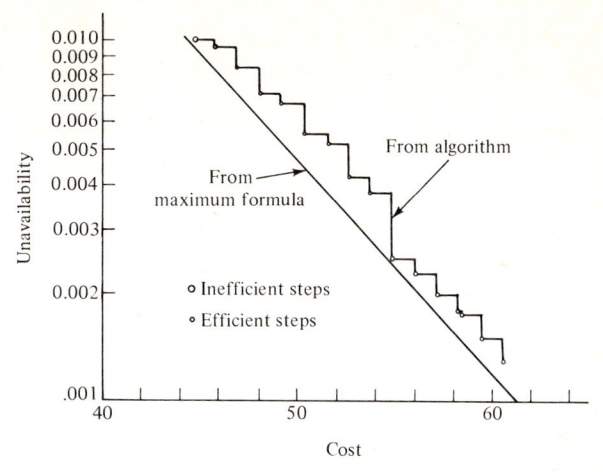

FIGURE 19-3
Availability as a function of cost for sample system.

in availability with a relatively large increase in cost). These inefficient steps are often eliminated at the next level of computation.

The dominating sequence for the system is given in Table 19-9. The accompanying availability-cost function is shown in Fig. 19-3.   ////

Table 19-9 DOMINATING SEQUENCE FOR STAGES 1, 2, 3, 4

	Cost	Unavailability	Stage 1	Stage 2	Stage 3	Stage 4
0	44.6	0.0100	5	5	4	3
1	45.7	0.0096	4	6	4	3
2	46.8	0.0084	4	5	5	3
3	48.0	0.0071	5	5	5	3
4	49.1	0.0067	4	6	5	3
5	50.3	0.0055	5	6	5	3
6	51.5	0.0052	6	6	5	3
7	52.5	0.0042	5	5	5	4
8	53.6	0.0038	4	6	5	4
9	54.8	0.0025	5	6	5	4
10	56.0	0.0023	6	6	5	4
11	57.1	0.0020	5	7	5	4
12	58.2	0.0018	5	6	6	4
13	58.3	0.00176	6	7	5	4
14	59.4	0.0015	6	6	6	4
15	60.5	0.0013	5	7	6	4

20

SOME PROBLEMS TO WRESTLE WITH

The problem statements given in this chapter are abstracted from situations presented in published literature. They give an opportunity to develop solutions to a variety of real problems.

20-1 GAS TURBINE ENGINE RELIABILITY

Dispatch reliability of gas turbine engines on aircraft is an important reliability parameter. It is the most obvious engine parameter and is reflected in terms of flight delays or cancellations (commercial airlines) or additional aircraft or engines to meet availability requirements (military). Analysis of performance data indicates that failures in engine controls constitute a significant fraction of the failures. A typical engine controller provides control of thrust, acceleration, deceleration, compressor stator vane angles, and bleeding. These functions are performed by a self-contained hydromechanical unit mounted on the engine. This unit senses speed, temperature, and pressures; positions cams, levers, links, and servomechanisms; and meters fuel flow. Improvement in control reliability by redundancy of hydromechanical units is not feasible because this would result in too great an increase in engine weight and cost. Devise a system for improving controller reliability, preferably involving some redundancy.

20-2 DESIGN ADEQUACY ASSESSMENT

Nuclear power generating stations are designed and constructed under strict regulations and quality-control standards to assure safety. A highly reliable control system is required for normal operation. Plant design also includes a dependable protection system to prevent small incidents from becoming large ones. A third level of defense equips nuclear stations with a variety of safeguard systems which are intended to mitigate post-accident consequences. These safeguard systems are verified by analyses of the system under assumed sets of accident conditions. Such analyses, however, are not sufficient, since safeguard systems must not only be capable of proper performance but must be available to perform. Develop a method of assessing reliability which also assesses availability of a system to perform its intended function.

20-3 ACCELERATED FATIGUE TEST

Fatigue life of components, e.g., frame-and-body systems in automobiles, subjected to varying magnitude of loading, is often estimated by applying Miner's rule, which combines a fatigue (S-N) curve and frequency of stress occurrence. This procedure has two disadvantages: (1) determination of the S-N curve usually requires several specimens and extended testing, and (2) considerable doubt exists as to the validity of relationships such as Miner's rule.

    To overcome these, laboratory simulations have been developed. Fatigue life can be estimated rather accurately if the actual random-stress wave is applied without modification. Unfortunately, this requires excessive testing time. As a consequence, accelerated simulation tests have been developed. An overstress testing procedure [Sec. 9-4-2 and Eq. (9-17)] is sometimes used, but too great an overstress can result in a different location and "process" of failure. A programmed fatigue test composed of steplike stress sequences does not require additional expensive testing equipment and substantially reduces testing time. At the same time, however, selection of a stress-frequency counting method is difficult for a wide-band random-stress wave, the effect of understressing is uncertain, and the actual stress-cycle sequence is ignored. Develop a laboratory simulation method for accelerated fatigue testing which overcomes the disadvantages cited above.

20-4 SNAP-FIT ASSEMBLIES IN POLYMERS

Many electronic appliances have rods with a push button attached to the end. The reliability requirement is that the push button should stay attached to the rod during the lifetime of the device, but it should also be possible to remove

198 INTRODUCTION TO RELIABILITY IN DESIGN

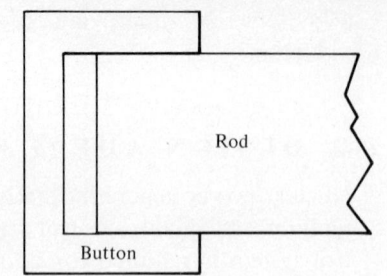

FIGURE 20-1
Schematic rod and push-button assembly.

the button without impairing the function of the component attached to the rod. A great variety of arrangements have been tried, with some proving rather unsatisfactory. A schematic arrangement is shown in Fig. 20-1. Design an appropriate configuration and develop a probabilistic design technique for determining dimensions and tolerances for a snap-fit assembly to meet the given requirements.

20-5 STEERING KNUCKLE REDESIGN

In Sec. 1-4, the second task of reliability engineering is stated as finding the best way of increasing reliability. The first two methods listed are obviously limited by the current state of technology. The balance of the first ten chapters implicitly assumes the limits of technology have been achieved. This assump-

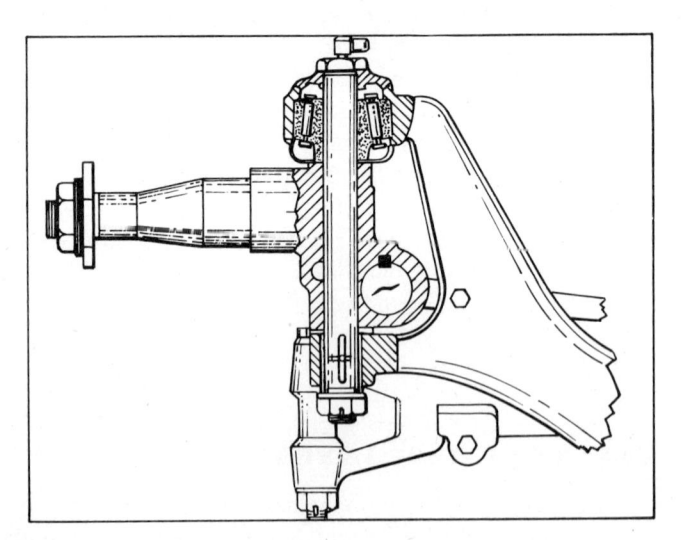

FIGURE 20-2
An early knuckle on a solid transverse axle (circa 1915).

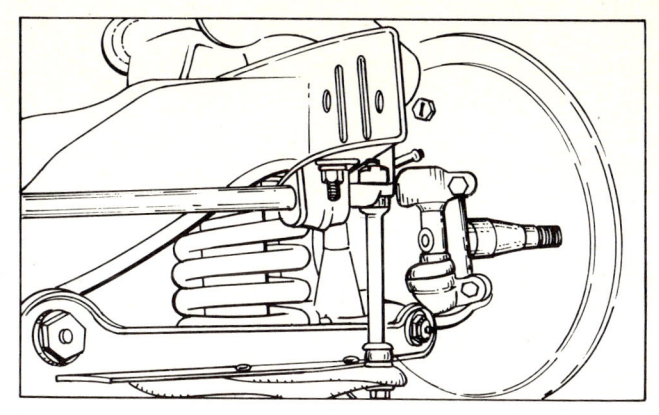

FIGURE 20-3
Knuckle, support, and steering pin for an independent suspension.

tion is, of course, not always valid. For example, consider the case of the steering knuckle on the Cadillac.

The earliest type of steering knuckle used on a straight-beam axle had an assembly consisting of a spindle-knuckle, bearings, and steering arm (Fig. 20-2). When the independent front suspension was introduced, the system was modified to knuckle support, knuckle-spindle, bearing, and steering arm (Fig. 20-3). Further modification to include a ball-joint front suspension permitted some simplification but still required assembly of a knuckle-spindle and steering arm (Fig. 20-4). These assemblies, in turn, were combined with a disk-brake mounting bracket (Fig. 20-5).

Cadillac's market includes ambulances, limousines, and other vehicles

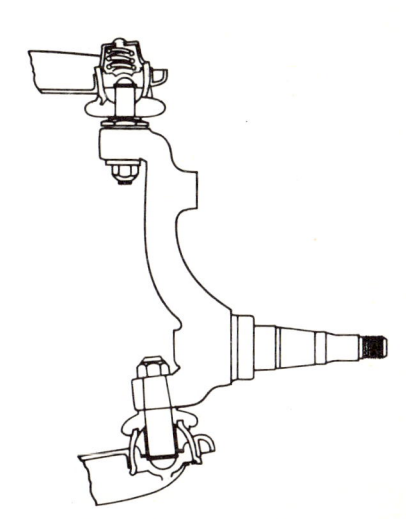

FIGURE 20-4
A steering knuckle for a ball-joint suspension.

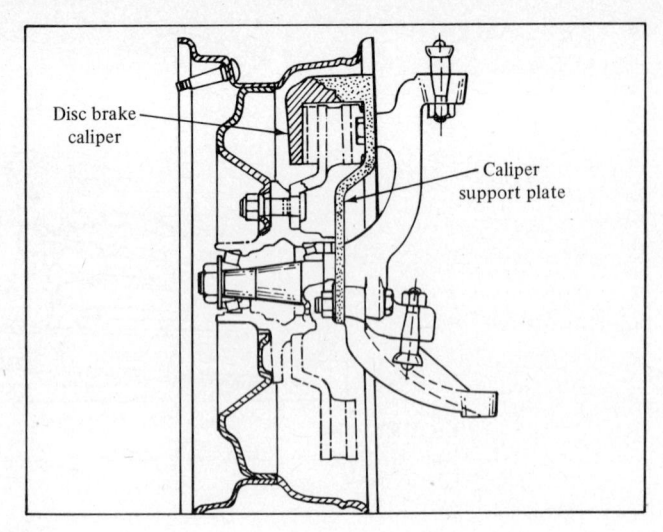

FIGURE 20-5
A steering knuckle and ball-joint assembly for disk brakes.

which are substantially heavier than many vehicles. Suspensions are often subjected to more severe loading than the common passenger car. With the standard automobile, the knuckle and spindle are commonly made from a single forging. With a higher load requirement, Cadillac made the spindle and knuckle separately and then assembled the two by shrink-fitting the spindle into the knuckle, since a more desirable stress distribution was achieved in this way. Machining operations on the spindle were also simplified. The resulting system included a knuckle, spindle, brake mount, and steering arm; i.e., four components which had to be assembled by humans, with resulting variation in the assembly and a product that was less than completely reliable.

What would you do to obtain a better assembly with less variability and thus, by implication, greater reliability?

20-6 STRUCTURAL CRITERIA FOR COMPOSITE MATERIALS

Composite materials technology has moved from characterization of materials to component development and usage. Nonetheless, many companies are reluctant to use composites because of (1) lack of extensive experience, (2) tendency of many composites to fail at relatively low strains, (3) variability in properties, (4) higher cost, (5) potential degradation of epoxies, and (6) uncertain development costs. Use of composites can, however, improve performance and effectiveness of structures. A composite airframe, for example, is desirable, but high confidence in composite technology must be generated. Design a plan involving design practices and structural design criteria which will consolidate existing technology and guide future composite materials research.

20-7 RELIABILITY IN SOFTWARE SYSTEMS

Software reliability can be defined as the probability that the application program (with the accompanying operating system, data base, and computing environment) will perform the intended functions at the time those functions are required. Testing of software provides tangible and repeatable proof that the program performs as intended. Such testing, however, must be within cost, schedule, and performance constraints. Few large real-time programs have ever been completely tested in the sense that every logic data path has been successfully executed for all possible options. While management seeks to test every logic path at some level at least once with a numerical check, this is rarely possible in practice. Thus the customer must be willing to accept, and the contractor willing to release, a program that is not completely checked but is shown to be reasonably free of errors. Devise a system or technique for achieving software reliability within these constraints.

20-8 SERVICE LIFE EVALUATION

Service life of many electronic systems includes periods of storage, dormancy, and cyclic operations. This situation presents a number of questions to equipment designers:

1 What are the effects of storage, dormancy, and power on-off cycling on equipment reliability? How significant are these effects?
2 Which will cause more degradation: continuous power (even at low levels) or switch-on and switch-off?
3 What is the proper balance of on-off cycling and operating at design level for minimum failure occurrence?
4 How can the number of field failures be minimized?

Develop a system or technique for evaluating the combined effects of storage, dormancy, on-off cycling (at various levels), and operation to provide the best design approach under constraints of cost and operational requirements.

20-9 DEPLOYMENT OF A SPACE ANTENNA

A space antenna has a number of rigid ribs, shaped like parabolas, mounted about a central hub with a mechanical deployment device. As stowed, the tips of the ribs are held in preloaded devices restrained by a cable. Upon command to deploy, the cable is severed by cutters. The preload force overcomes the restraining force and the ribs "pop" free. Deployment continues through a ball-screw and carrier mechanism operated by torque motors until deployment is complete. A radio frequency (RF) reflective surface of flexible, metallic double mesh is attached to the ribs and pulled tight during deployment and latching. Estimate the reliability of a one-shot space antenna if the criteria are full deployment and latching.

20-10 RELIABILITY AND ESP

All reliability problems can be considered as design-caused or people-caused. If design is perfect, an inherent reliability of 100 percent is theoretically possible. Operations and maintenance, both people-associated, can only degrade reliability. Either of these activities can be regarded as improved if people can be prevented from repeating previous mistakes (theirs or others). Develop a system which will prevent (or minimize) repetition of design mistakes. This system should supplement, rather than substitute for, known design control measures such as design standards, design review checklists, and design manuals.

20-11 PLASTIC INTEGRATED CIRCUITS

An integrated circuit (IC) package serves three functions:

1 It provides protection for the chip against mechanical abuse and the chemical environment.
2 It provides electrical connections for the chip.
3 It provides a convenient means for handling tiny IC chips.

Hermetically sealed ICs have high reliability as a result of providing the chip with excellent protection from its environment. Plastic packaged ICs, however, are more susceptible to permeability from moisture, low compressive strength, thermal instability, and lack of compatibility at interfaces of device components. Design a method for developing a highly stable, well-controlled ambience around the chip for environmental protection and develop a system, technique, or set of guidelines for determining suitability of plastic packaged ICs.

BIBLIOGRAPHY

General

G-1 ABRAHAM, L. H.: "Structural Design of Missiles and Space Craft," McGraw-Hill Book Company, New York, 1962.

G-2 AGREE Report: *Reliability of Military Electronic Equipment,* Office of the Assistant Secretary of Defense (Research and Engineering), Washington, D.C., 1957.

G-3 AMSTADTER, B. L.: "Reliability Mathematics," McGraw-Hill Book Company, New York, 1971.

G-4 ANKENBRANDT, F. L.: "Maintainability Design," Engineering Publishers, Division of A C Book Company, Inc., Elizabeth, N.J., 1963.

G-5 ARINC Research Technical Staff: "Reliability Engineering," Prentice-Hall, Inc., Englewood Cliffs, N.J., 1964.

G-6 BARLOW, R. E., and F. PROSCHAN: "Mathematical Theory of Reliability," John Wiley & Sons, Inc., New York, 1965.

G-7 BARLOW, W. R., L. HUNTER, and F. PROSCHAN: "Probabilistic Models in Reliability Theory," John Wiley & Sons, Inc., New York, 1962.

G-8 BAZOVSKY, I: "Reliability Theory and Practice," Prentice Hall, Inc., Englewood Cliffs, N.J., 1961.

G-9 BECKMANN, P: "Probability in Communication Engineering," Harcourt, Brace & World, Inc., New York, 1967.

G-10 BREIPOHL, A. M.: "Probabilistic Systems Analysis," John Wiley & Sons, Inc., New York, 1970.

G-11 CALABRO, S. R.: "Reliability Principles and Practices," McGraw-Hill Book Company, New York, 1962.

G-12 CHORAFAS, D. N.: "Statistical Processes and Reliability Engineering," Van Nostrand Company, Inc., Princeton, N.J., 1960.

G-13 COX, D. R.: "Renewal Theory," John Wiley and Sons, Inc., New York, 1962.

G-14 DUMMER, G. W. A., and N. B. GRIFFIN: "Electronics Reliability—Calculation and Design," Pergamon Press, New York, 1966.

G-15 EPSTEIN, B.: "Quality Control and Reliability Handbook H 108," Government Printing Office, Washington, D.C., 1960 (Sampling Procedures and Tables for Life and Reliability Testing).

G-16 GOODE, H. P., and J. H. KAO: "Quality Control and Reliability Technical Report No. TR-3," Office of Assistant Secretary of Defense (Installations and Logistics), Washington, D.C., 1961 (Sampling Procedures and Tables for Life and Reliability Testing Based on the Weibull Distribution).

G-17 GOLDMAN, A. S., and T. B. SLATTERY: "Maintainability: A Major Element of System Effectiveness," John Wiley and Sons, Inc., New York, 1964.

G-18 GRYNA, F. M., N. J. RYERSON, and S. SWERLING (eds.): "Reliability Training Text, 2d ed.," 1 East 79th Street, Institute of Radio Engineers, New York, 1960.

G-19 GRYNA, F. M., JR. (ed.): Reliability Theory and Practice. Sixth Annual Workshop, Chicago, Ill., June 22–24, 1965, Sponsored by Electrical Engineering Division of ASEE at the Illinois Institute of Technology (available from ASEE Headquarters).

G-20 HAVILAND, R. P.: "Engineering Reliability and Long Life Design," D. Van Nostrand Company, Inc., Princeton, N.J., 1964.

G-21 HENNEY, K. (ed.): "Reliability Factors for Ground Equipment," McGraw-Hill Book Company, New York, 1956.

G-22 IRESON, W. G. (ed.): "Reliability Handbook," McGraw-Hill Book Company, New York, 1966.

G-23 KENNEY, D. P.: "Application of Reliability Techniques," Argyle Publishing Corporation, New York, 1966. (A self-instructional programmed training course book.)

G-24 LANDERS, R: "Reliability and Product Assurance," Prentice-Hall, Inc., Englewood Cliffs, N.J., 1963.

G-25 LEAKE, C. E.: "Understanding Reliability," United Testing Laboratory, Pasadena, Cal., 1960.
G-26 LIN, Y. K.: "Probabilistic Theory of Structural Dynamics," McGraw-Hill Book Company, New York, 1967.
G-27 LIPSON, C., J. KERAWALLA, and L. MITCHELL: "Engineering Applications of Reliability," University of Michigan, Ann Arbor, 1963.
G-28 LLOYD, D. K., and M. LIPOW: "Reliability, Management and Mathematics," Prentice-Hall, Inc., Englewood Cliffs, N.J., 1962.
G-29 Mechanical Engineering series: Reliability and Maintainability, February–June (inclusive), 1966.
G-30 MULLIGAN, T.: Reliability Predictions from Test Statistics, 22nd Annual Technical Conference, American Society of Body Engineers, 1967.
G-31 MYERS, R. H., K. L. WONG, and H. M. GORDY, (eds.): "Reliability Engineering for Electronic Systems," John Wiley and Sons, Inc., New York, 1964.
G-32 PIERUSHKA, E.: "Principles of Reliability," Prentice-Hall, Inc., Englewood Cliffs, N.J., 1963.
G-33 PURSER, P. E., M. A. FAGET, and N. F. SMITH: "Manned Spacecraft: Engineering Design and Operation," Fairchild Publications, Inc., Book Div., New York, 1964.
G-34 *Reliability and Maintainability Symposia Proceedings* (known by different names starting in 1954, annually since 1956) Institute of Electrical and Electronic Engineers, New York.
G-35 ROBERTS N. H.: "Mathematical Methods in Reliability Engineering," McGraw-Hill Book Company, New York, 1964.
G-36 Rome Air Development Center: Reliability Notebook Supplement 1, Section 8 (AB 161894-1), U.S. Dept. of Commerce, Office of Technical Services, Washington, D.C., September 1, 1960.
G-37 ROSS, L. R.: Why Manufacturing Can't Build It Like the Print, 22nd Annual Technical Conference, American Society of Body Engineers, October, 1967.
G-38 *Reliability Control in Aerospace Equipment Development,* Society of Automotive Engineers, Inc., SAE Technical Progress Series, Vol. 4, 1963.
G-39 SANDLER, G. H.: "System Reliability Engineering," Prentice-Hall, Inc., Englewood Cliffs, N.J., 1963.
G-40 SHWOP, J., and H. SULLIVAN: "Semiconductor Reliability," Elizabeth, N.J., 1961.
G-41 TANGERMAN, E. J. (ed.): "Manual of Reliability," *Product Engineering,* Reader Service Department, New York, 1960.
G-42 TRIBUS, M. "Rational Description, Decisions and Designs," Pergamon Press, New York, 1969.
G-43 WEYMUELLER, C. R.: Making Products That Last—The Story of the Westinghouse Reliability Program, *Metal Progress,* vol. 88, no. 5, pp. 98–102, 108, 112, 114, 116, 118, 1965.
G-44 WILCOX, M. H., and W. C. MANN (cds.): "Redundancy Techniques for Computing Systems," Spartan Press, Washington, D.C., 1962.
G-45 ZELEN, M. (ed.): "Statistical Theory of Reliability," The University of Wisconsin Press, Madison, 1962.

Chapter 8

8-1 O'CONNELL, E. P. (ed.): "Handbook of Product Maintainability," Reliability Division, American Society for Quality Control, Aug., 1973.

Chapter 9

9-1 MIL-STD-105D, Sampling Procedures and Tables for Inspection by Attributes.
9-2 MIL-M—8555A, Missiles, Guided: Design and Construction, General Specification for.
9-3 MIL R-38100A (USAF), Reliability and Quality Assurance Requirements for Established Reliability Parts, General Specifications for.

9-4   MIL-S-38103A (USAF), Semiconductor Device, Established Reliability, General Specification for.
9-5   MIL-STD-810A (USAF), Environmental Test Methods for Aerospace and Ground Equipment.
9-6   MIL-STD-781A, Reliability Tests, Exponential Distribution.
9-7   Sampling Procedures and Tables for Life and Reliability Testing, Handbook H108, U.S. Government Printing Office, 1960.
9-8   DUNCAN, A. J.: "Quality Control and Industrial Statistics," Richard D. Quinn, Inc., Homewood, Ill., 1965.
9-9   BOWKER, A. H., and G. J. LIEBERMAN: "Engineering Statistics," Prentice-Hall, Inc., Englewood Cliffs, N.J., 1959.
9-10 CONOVER, J. C., H. R. JAECKEL, and W. J. KIPPOLA: Simulation of Field Loading in Fatigue Testing, SAE Paper 660102, January, 1966.
9-11 Factorial Experiments
   Davies, O. L.: "Design and Analysis of Industrial Experiments," Hafner Publishing Company, Inc., N.Y., 1967.
   Hicks, C. R.: "Fundamental Concepts in the Design of Experiments," Holt, Rinehart and Winston, Inc., New York, 1964.
   Li, C. C.: "Introduction to Experimental Statistics," McGraw-Hill Book Company, New York, 1964.
   Peng, K. C.: "Design & Analysis of Scientific Experiments," Addison Wesley Publishing Company, Inc., Reading, Mass., 1967.
9-12 ZELEN, M., Factorial Experiments in Life Testing, *Technometrics,* vol. 1, no. 3, August 1959.

Chapter 10

10-1 BROWN, P. I., and R. J. O'BRIEN: Considering Liability in Product Design, *Ind. Res.,* pp. 28–31, December, 1973.
10-2 JACOBS, W: Products-Liability Suits and the Engineer, *Mech. Eng.,* pp. 12–19, November, 1972.
10-3 JORDAN, W. E.: Failure Modes, Effects and Criticality Analyses, Proceedings, 1972 Annual Reliability and Maintainability Symposium, 25–27 January 1972, pp. 30–37.
10-4 MILES, L. D.: "Techniques of Value Analysis and Engineering," McGraw-Hill Book Company, New York, 1961.
10-5 NORQUIST, W. E.: Steps companies can take to increase the reliability of consumer products, Paper 72-DE-8, Design Engineering Conference, Chicago, 8–11 May 1972, ASME, New York.

Chapter 11

11-1 JENSEN, P. A., and M. BELLMORE: An Algorithm to Determine the Reliability of a Complex System, *IEEE Trans. on Reliability,* vol. R-18, no. 4, pp. 169–174, November 1969.

Chapter 12

12-1 DAGEN, H.: Preliminary Report on the Relationship Between Predicted Failure Rates and Observed Failure Rates, *Ann. Reliability Maintainability,* vol. 5, American Institute of Aeronautics and Astronautics, New York, pp. 704–709, 1966.
12-2 TANNER, T. L.: Operational Reliability of Components in Selected Systems, 1966 Symposium on Reliability.
12-3 WAGNER, R. H., and E. H. PURVIS: Electronic Design—Calculated versus Observed Reliability, *Ann. Reliability Maintainability,* vol. 4, Spartan Books Inc., Washington, D.C., pp. 375–382, 1965.

Chapter 13

13-1 BAILEY, FREDERICK J., and JOHN C. FRENCH: Reliability and Flight Safety, Mercury Project Summary, NASA SP-45, Washington, D.C., 1963.
13-2 DOSHAY, I.: Reliable Spacecraft Designs, SAE Paper 343B, New York, 1961.
13-3 MORRIS, OWEN G.: *Apollo Reliability Analysis, Astronaut. Aeron.,* vol., 3 no. 4, American Institute of Aeronautics and Astronautics, Easton, Pa., April 1965.
13-4 Surveyor Project Staff: Project Description and Performance, Surveyor Project Final Report, vol. 1, Technical Report 32-1265, Jet Propulsion Laboratory, July 1, 1969.
13-5 Where Components Must Be Improved, *Missiles Rockets,* vol. 12, no. 24, American Aviation Publications, Washington, D.C., June 17, 1963.

Chapter 14

14-1 BIRCHENOUGH, A. G., and F. GOURASH: Design and Performance Analysis of a Medium-Power DC-DC Converter, NASA Technical Note, NASA TN D-5643, National Aeronautics and Space Administration, Washington, D.C., February 1970.

Chapter 15

15-1 CARPENTER, R. B., JR.: Reliability for Manned Interplanetary Travel, *Ann. Reliability Maintainability,* vol. 4, pp. 413–419, 1965.
15-2 CARPENTER, R. B., JR.: Space: Manned Interplanetary Travel, *Mech. Eng.* pp. 44–48, June, 1966.
15-3 CARPENTER, R. B., JR.: Safety Assurance for Extended Manned Missions, *J. Spacecraft Rockets,* vol. 4, no. 4, pp. 448–451, April, 1967.
15.4 BURROWS, R. W.: The Long-Life Spacecraft Problem, *J. Spacecraft Rockets,* vol. 5, no. 3, p. 362, March, 1968.
15-5 MILLER, J. E., and J. FELDMAN: Gyro Reliability in the Apollo Guidance, Navigation, and Control System, *J. Spacecraft Rockets,* vol. 5, no. 6, pp. 638–643, June, 1968.

Chapter 16

16-1 PONTECORVO, A. B.: A Method of Predicting Human Reliability, *Ann. Reliability Maintainability,* vol. 4, pp. 337–342, 1965.
16-2 BARONE, M. A.: A Methodology to Analyze and Evaluate Critical Human Performance, Annals of Reliability and Maintainability, vol. 5, pp. 116–122, 1966.
16-3 BROWN, E. S: System Safety and Human Factors: Some Necessary Relationships, "Proceedings, 1974 Reliability and Maintainability Symposium," IEEE, New York, pp. 197–200.
16-4 CELINSKI, O., and M. MASTER: An Activity Model for Predicting the Reliability of Human Performance, *Ibid.,* pp. 340–348.
16-5 HUSTON, R. L., and A. M. STRAUSS: Human Reliability in Man-Machine Interactions, *op. cit.,* pp. 329–334.
16-6 KELLY, C. W., and S. BARCLAY: Improvement of Human Reliability Using Bayesian Hierarchal Inference, *op. cit.,* pp. 320–328.
16-7 REGULINSKI, T. L.: Human Performance Reliability Modeling in Time Continuous Domain, Air Force Institute of Technology, Wright Patterson Air Force Base, Ohio, Jan., 1974.
16-8 THOMPSON, C. W. N., Model of Human Performance Reliability in Health Care Systems, ibid. pp. 335–339.

Chapter 17

17-1 BALL, M., and F. HARDIE: "Redundancy for Better Maintenance of Computer Systems," *Comput. Design,* vol. 8, no. 1, pp. 50–53, January, 1969.

17-2 HSIAO, M-Y., and J. T. TOU: "Application of Error-Correcting Codes in Computer Reliability Studies," *IEEE Trans.* on Reliability, vol. R-18, no. 3, pp. 108–118, August, 1969.

17-3 RAO, T. R. N.: "Use of Error Correcting Codes on Memory Words for Improved Reliability," *IEEE Trans* on Reliability, vol. R-17, no. 2, pp. 91–96, June, 1968.

17-4 ROBERTS, D. C.: "Increasing Reliability of Digital Computers," Computer Design, vol. 8, no. 1, pp. 44–48, January, 1969.

Chapter 18

18-1 HAUGEN, E. B.: "Probabilistic Approaches to Design," John Wiley & Sons, Inc., New York, 1968.

18-2 BAKER, H. C.: Reliability Analysis and Test of Aluminum Honeycomb Structures, *Ann. of Reliability Maintainability,* vol. 4, pp. 431–436, 1965.

18-3 CORNELL, A. C.: A Probability-Based Structural Code, American Concrete Institute, Fall Convention, Memphis, Tenn., November, 1968.

18-4 DAO-THIEN, M., and M. MASSOUD: On the Relation Between the Factor of Safety and Reliability, ASME Paper No. 73-WA/De-1.

18-5 MISCHKE, C. R.: A Rationale for Mechanical Design to a Reliability Specification *Proc.,* ASME Design Technology Transfer Conference, New York, October 5–9, 1974, pp. 221–233.

18-6 MISCHKE, C. R.: "Implementing Mechanical Design to a Reliability Specification," *Ibid.* pp. 235–248.

18-7 MITTENBERGS, A. A.: The Materials Problem in Structural Reliability, *Annals of Reliability and Maintainability,* vol. 5, pp. 148–158, 1966.

18-8 SMITH, C. O.: Design of Circular Members in Bending and Torsion to Probabilistic Criteria, Fourth National Conference on Engineering Design, Dartmouth College, July 1967.

18-9 SMITH, C. O., and E. B. HAUGEN: "Design of Circular Members in Torsion to Probabilistic Criteria," *Design News,* pp. 108–115 (1 col/p), November 8, 1968.

18-10 SMITH, C. O.: Structural Designs Based on Probability, *Design News,* pp. 98–108 (1 col/p), December 8, 1969.

18-11 SMITH, C. O.: Design of Pressure Vessels to Probabilistic Criteria, First International Conference on Structural Mechanics in Reactor Technology, Berlin, Germany, September 1971.

18-12 SMITH, C. O: Design Relationships and Failure Theories in Probabilistic Form, *Nucl. Eng. Design,* vol. 27, no. 2, pp. 286–292, May 1974.

Chapter 19

19-1 JENNY, J. S., and W. F. LAPSON: "Feasibility Study of an Automated, Unmanned Geophysical Observatory for Operation in Antarctic," Stanford Electronics Laboratories (SU-SEL-68-010), February, 1968.

19-2 SITES, M. J.: Stanford University, Stanford, California, Private Communication, February 20, 1975.

19-3 KETTELLE, J. D., JR.: Least-Cost Allocations of Reliability Investment, *Operations Research,* vol. 10, pp. 249–265, March-April 1962.

19-4 TILLMAN, F. A.: Optimization by Integer Programming of Constrained Reliability Problems with Several Modes of Failure, *IEEE Transactions on Reliability,* May 1969.

19-5 LIANG, T. F., et. al.: Optimization of System Reliability, *IEEE Transactions on Reliability,* September 1963.

19-6 BARLOW, R. E., and L. C. HUNTER: "Criteria for Determining Optimum Redundancy," *IRE Trans. Reliability Quality Control,* April 1960.

APPENDIX A
NOTES ON PROBABILITY

A-1 INTRODUCTION

The theory of probability is that part of mathematics which describes statistical phenomena and is fundamental to the techniques of statistical inference. Probability and statistics are so interrelated that it is very difficult to discuss one without some understanding of the other, but it must be recognized that they are not synonymous terms. A knowledge of probability makes it possible to interpret statistical results, since a large number of statistical procedures involve conclusions based on samples which are affected by random variation. Probability theory allows us to numerically express the inevitable uncertainties in the resulting inferences or conclusions.

One basic thought is pertinent here, as in all engineering endeavors, as expressed by J. Neyman:[†]

> Whenever we use mathematics in order to study some observational phenomena we must essentially begin by building a mathematical model (deterministic or probabilistic) for these phenomena. Of necessity, the model must simplify matters and certain details must be ignored. The success of the model depends on whether or not the details ignored

[†] University of California Publications in Statistics, Vol. I, University of California Press, Berkeley, 1954.

are really unimportant in the development of the phenomena studied. The solution of the mathematical problem may be correct and yet be in considerable disagreement with the observed data simply because the underlying assumptions made are not warranted. It is usually quite difficult to state with certainty, whether or not a given mathematical model is adequate *before* some observational data are obtained. In order to check the validity of a model, we must *deduce* a number of consequences of our model and then compare these *predicted* results with observations.

We have no influence over what we observe, but we should use our critical judgment in choosing the model we use.

A-2 DEFINITIONS

Statistics is the science dealing with the collection, organization, analysis, and interpretation of numerical data.

Probability is the study of random or nondeterministic experiments.

A-3 CLASSICAL DEFINITION OF PROBABILITY

Assume an event E can occur in n ways out of a total of N possible equally likely ways. Then the probability of occurrence of the event (usually called success) is defined by

$$p = \Pr\{E\} = \frac{n}{N} \qquad \text{(A-1)}$$

The probability of nonoccurrence of the event (usually called failure) is defined by

$$q = \Pr\{\text{not } E\} = \frac{N-n}{N} = 1 - \frac{n}{N} = 1 - \Pr\{E\} \qquad \text{(A-2)}$$

Thus $p + q = 1$ or $\Pr\{E\} + \Pr\{\text{not } E\} = 1$.

A-4 RELATIVE FREQUENCY DEFINITION OF PROBABILITY

It may be noted that the classical definition contains the words "equally likely." Since these words appear to be synonymous with "equally probable," the definition is circular. For this reason, a statistical definition of probability has been urged by many. In such a definition, the estimated probability or empirical probability of an event is taken as the relative frequency of occurrence of the event when the number of observations is very large. The probability itself is the limit of the relative frequency as the number of observations increases indefinitely.

It should be emphasized that within the definition of probability, we should consider not only the nature of the event but also the nature of the category of observations. This is demonstrated by Bertrand's paradox. The observation consists of randomly drawing a chord in a circle (radius = R) and reading its length. The probability required is that the length of the chord is less than the side of the inscribed equilateral triangle, i.e., a length less than $R\sqrt{3}$. The process of "randomly drawing a chord" is not well defined, thus there may be many contradictory answers. Consider two possibilities: (1) The perpendicular distance from the center of the circle to the chord lies between O and R. The chord is less than $R\sqrt{3}$ if this distance is greater than $R/2$. If the probability is measured as the ratio of the favorable length $R/2$ to the possible length R, the desired probability is $\frac{1}{2}$. (2) The central angle of the chord is a number between 0 and π. The chord is less than $R\sqrt{3}$ if the central angle is less than $2\pi/3$. If the probability is measured as the ratio of the favorable central angle $2\pi/3$ to the possible angle π, the desired probability is $\frac{2}{3}$.

Probability plays the role in statistics of a substitute for certainty. In studying and using probability, there are essentially three kinds of questions:

1 The question of what we mean when we say that a probability is 0.80, 0.91, etc. The answer is essentially defined in terms of relative frequency.
2 The question of how to obtain numerical values of probabilities. The answer is found by solving the problem of estimation—estimations are usually made by observing the relative frequency with which similar events have occurred in the past.
3 The question of how to use known probabilities to calculate others. The answer is the calculation of probabilities of relatively complex events in terms of known (or assumed) values of probabilities of simpler kinds of events.

EXAMPLE A-1 An honest coin is tossed twice (or two coins are tossed simultaneously). What is the probability of getting *one head?*

One analysis might be: In a given toss, there might be 0, 1, or 2 heads. One might deduce that the probability of *one head* is $\frac{1}{3}$. This is incorrect since all the outcomes listed are not equally likely.

Consider all the equally likely possibilities: HH, HT, TH, TT. The correct probability of getting one head is $\frac{1}{2}$. ////

EXAMPLE A-2 If two dice are thrown once, what is the probability of getting (*a*) a total of 9; or (*b*) a total different from 9?
(*a*) A total of 9 can be obtained by $6 + 3$, $3 + 6$, $4 + 5$, or $5 + 4$, where the first number is shown by the first die and the second number by the second die. There are 36 equally likely different ways in which two dice can fall. There are 4 ways of getting a total of 9. The probability of getting a total of 9 is

$$p(\text{sum} = 9) = \tfrac{4}{36} = \tfrac{1}{9}$$

(b) Since the dice must total 9 or a number different from 9, the probability of getting a total other than 9 is

$$p(\text{sum} \neq 9) = 1 - \tfrac{1}{9} = \tfrac{8}{9}$$

This result could also have been found by counting in the same way as the answer to part (a).   ////

A-5 FUNDAMENTAL LAWS OF PROBABILITY

For a single event, there is a variety of ways in which we may choose among infinitely many numbers. When we are concerned with simultaneous probabilities for several events, we must introduce these according to certain relationships between the corresponding relative frequencies, since the same relationships must hold true for all probabilities. This requires the statement of fundamental laws which form the bases of all the deductions which follow but which themselves can not be proved mathematically. (This is analogous to the axioms in Euclidean geometry, Newton's laws in mechanics, and Maxwell's equations in electrodynamics.)

A relative frequency is always of the form $f = n_i/N$, where $0 \leq n_i \leq N$, and thus $0 \leq f \leq 1$. The same relationship must hold for any probability.

First axiom: The value of a probability P is a number between 0 and 1, both limits included.

$$0 \leq P \leq 1 \quad (\text{A-3})$$

By a *certain event* E we understand an event which occurs in every observation.

Second axiom: The probability of a certain event is 1.

$$P(E) = 1 \quad \Pr\{E\} = 1 \quad (\text{A-4})$$

The converse need not be true. If an event has a probability of 1, this means only that it is *practically certain,* i.e., its relative frequency is very close to 1 if the number of observations is large.

By an *impossible event* 0 we understand an event which does not occur in any observation.

Third axiom: The probability of an impossible event is 0.

$$P(0) = 0 \quad \Pr\{0\} = 0 \quad (\text{A-5})$$

As in Eq. (A-4), the converse need not be true. If an event has a probability of 0, this means only that it is *practically impossible,* i.e., its relative frequency is very close to 0 if the number of observations is large.

Equations (A-4) and (A-5) provide an illustration of the statement that we must not confuse reality with the model, for a theory can prove only statements about the model, not about reality.

Consider two events A and B in a series of N observations. In each observation one and only one of four possibilities may occur:

1. A has occurred, but not B. (n_1)
2. B has occurred, but not A. (n_2)
3. Both A and B have occurred. (n_3)
4. Neither A nor B has occurred. (n_4)

$$n_1 + n_2 + n_3 + n_4 = N$$

Fourth axiom: The probability that at least one of two events occurs is equal to the sum of the probabilities of each event minus the probability of both events occurring simultaneously.

$$P(A + B) = P(A) + P(B) - P(AB). \qquad (A\text{-}6)$$

This axiom is called the *addition law of probability* or, often, the law of "*either, or.*"

A special case of Eq. (A-6) occurs if the two events exclude each other. By definition, two or more events are *mutually exclusive* if they are such that not more than one of them can occur in a single trial (or observation). Then, if $p_1, p_2, p_3, \ldots, p_k$ are the separate probabilities of the occurrence of k mutually exclusive events, the probability P that some *one* of these events will occur in a single trial is

$$P = p_1 + p_2 + p_3 + \cdots + p_k$$
$$P(A + B) = P(A) + P(B) \qquad (A\text{-}6a)$$

Fifth axiom: The probability of both of two events occurring is equal to the product of the absolute probability of one event and the conditional probability of the other.

$$P(AB) = P(A)P(B/A)$$
$$P(AB) = P(B)P(A/B) \qquad (A\text{-}7)$$

This axiom is called the *multiplication law* of probability or, often, the law of "*both, and.*" It is also called the law of *compound probabilities*.

Conditional probability, by definition, is the probability of the occurrence of event B, provided that event A has taken place (or is taking place or will, for certain, take place). This is the conditional probability of B relative to A and is written $P(B/A)$. If A and B are mutually exclusive, then $P(B/A)$ and $P(A/B)$ are both equal to zero.

A special case of Eq. (A-7) occurs if the two events are stochastically independent. By definition, two or more events are *independent* if the probability of occurrence (or nonoccurrence) of any one of them is in no way affected by the occurrence of any other. Otherwise, the events are *dependent*. If two events are mutually exclusive, they are dependent. If $p_1, p_2, p_3, \ldots, p_k$ are the separate probabilities of the occurrence of k independent events, the probability P that *all* of these events will occur in a single trial is

$$P = p_1 p_2 p_3 \cdots p_k$$

$$P(AB) = P(A)P(B) \qquad \text{(A-7}a\text{)}$$

In the case of k dependent events, if the probability of the occurrence of the first event is p_1, and if, after this event has occurred, the probability of a second event is p_2, and if, after the first and second events have occurred, the probability of the occurrence of a third event is p_3, and so on, the probability P that all events will occur in the specified order is

$$P = p_1 p_2 p_3 \cdots p_k \qquad \text{(A-8)}$$

Equation (A-7*a*) is often considered a special case of Eq. (A-8). In general, it is more convenient to think of these two equations as separate probability laws.

To clarify notation, we use $P(B/A)$ to indicate the probability of occurrence of event B, given that event A has occurred. We use $P(B)$ to indicate the probability of occurrence of event B when we do not know whether another event has occurred.

The probability $P(B)$ is, therefore, the probability of the occurrence of event B before any trials have been made; accordingly, it is called a *prior* or an *a priori* probability. At the same time, the probability $P(B/A)$ gives the probability of occurrence of event B after it is known that event A has occurred and accordingly is a *posterior* or an *a posteriori* probability.

In these terms, it is possible to rewrite Eq. (A-7) to give

$$P(B/A) = P(AB)/P(A) \qquad \text{(A-7}b\text{)}$$

where $P(AB)$ denotes the probability of occurrence of both events, A and B.

It is often of interest to calculate an a posteriori probability from given a priori probabilities. This is made possible by: if $B_1, B_2, \ldots, B_k$ are k mutually exclusive events, of which one must occur in a given trial, i.e., $P(B_1) + P(B_2) + \cdots + P(B_k) = 1$, and A is any event for which $P(A) \neq 0$, then the conditional probability $P(B_i/A)$ for any event B_i, given that event A has occurred, is given by

$$P(B_i/A) = \frac{P(B_i)P(A/B_i)}{P(B_1)P(A/B_1) + P(B_2)P(A/B_2) + \cdots + P(B_k)P(A/B_k)} \qquad \text{(A-9)}$$

This is known as Bayes' theorem, Bayes' law, or Bayes' formula. It is very useful, but its use is limited in many cases since the a priori probabilities are unknown.

EXAMPLE A-3   A card is drawn at random from a standard deck of 52 playing cards. It is not replaced. A second card is drawn. What is the probability that the first card is an ace and the second is a king?

The probability of getting an ace on the first draw is $\frac{4}{52} = \frac{1}{13}$. The probability of getting a king on the second draw is $\frac{4}{51}$. The probability of getting both is the product

$$p = (\tfrac{1}{13})(\tfrac{4}{51}) = \tfrac{4}{663} \qquad ////$$

EXAMPLE A-4 A beginning golfer has a probability of $\frac{1}{2}$ of making a good shot if he uses the correct club but only a probability of $\frac{1}{3}$ if he uses the wrong club. He is carrying five clubs. If he chooses a club at random and makes a shot, what is the probability that it is a good shot?

The successful events are:

Right club, good shot $\quad p = (\tfrac{1}{5})(\tfrac{1}{2}) = \tfrac{1}{10}$
Wrong club, good shot $\quad p = (\tfrac{4}{5})(\tfrac{1}{3}) = \tfrac{4}{15}$

Both cases give a good shot. Therefore, the probability of a good shot is the sum or $\tfrac{11}{30}$. ////

EXAMPLE A-5 One card is drawn from each of two separate, well-shuffled decks. What is the probability that one of the two cards is the king of hearts?

$$p = \frac{1}{52} + \frac{1}{52} - \frac{1}{(52)(52)} = \frac{103}{2704}$$

The third term is obtained from: there are (52)(52) ways in which a card from one deck can be matched with a card from the other deck. Of these $(52)^2$ possibilities, however, there is only one which gives both kings of hearts. ////

EXAMPLE A-6 Two cards are drawn in succession from the same deck without replacing the first. What is the probability of drawing a club on the first draw? What is the probability of drawing a club on the second draw, given that the first one was a club? What is the probability of both cards being clubs?

$$p(1 \text{ club}) = \tfrac{13}{52} = \tfrac{1}{4}$$

A club having been drawn, there are 12 clubs left in the 51 cards. The probability of now drawing a club is

$$p = \tfrac{12}{51}$$

Probability of both cards being clubs is the product

$$p = (\tfrac{1}{4})(\tfrac{12}{51}) = \tfrac{1}{17} \qquad ////$$

EXAMPLE A-7 Referring to Example A-4, if the golfer makes a good shot, what is the probability that he used the right club? Let A be the event of making a good shot, RC the event of selecting the right club, and WC the event of selecting the wrong club. Then, $p(RC) = \tfrac{1}{5}$, $p(WC) = \tfrac{4}{5}$, $p(A|RC) = \tfrac{1}{2}$, and $p(A|WC) = \tfrac{1}{3}$. Using

Bayes' theorem,

$$p(RC/A) = \frac{(\tfrac{1}{5})(\tfrac{1}{2})}{(\tfrac{1}{5})(\tfrac{1}{2}) + (\tfrac{4}{5})(\tfrac{1}{3})} = \tfrac{3}{11}$$

Thus, only three out of eleven good shots can be attributed to the use of the right club. ////

A-6 COMBINATIONS AND PERMUTATIONS

The problem of computing probabilities of events in finite sample spaces in which equal probabilities are assigned to the elements in any given sample space reduces to that of counting the elements which make up the events. Counting such elements is often simplified by using the rules for combinations and permutations which deal with the grouping and arrangement of objects. They are useful in calculating probabilities by determining the number of favorable cases as well as the total number of possible cases.

A combination is each of the sets which can be made by using all or part of a given collection of objects, without regard to the order of the objects in the set.

A permutation is each different ordering or arrangement of all or part of a set of objects.

For example, find (a) the number of combinations, and (b) the number of permutations of the letters A, B, C, D in sets of three.

(a) The four letters can be taken in sets of three, without regard to order, in the following ways: ABC, ABD, ACD, and BCD. Thus there are four such combinations, i.e., there are four combinations of four objects taken three at a time.

(b) If ordering is also considered, there are the following permutations of the four letters in sets of three: ABC, ACB, BAC, BCA, CAB, CBA, ABD, ADB, BAD, BDA, DAB, DBA, ACD, ADC, CAD, CDA, DAC, DCA, BCD, BDC, CBD, CDB, DBC, DCB. Thus there are 24 such permutations, i.e., there are 24 permutations of four objects taken three at a time.

The two definitions and the example above can be enlarged: A permutation of n different objects taken k at a time is an *arrangement* of k out of the n objects with attention given to the order of arrangement. The number of permutations of n objects taken k at a time is denoted by $_nP_k$, P_k^n, $P_{n,k}$, or $P(n,k)$ and is given by

$$P(n,k) = n(n-1)(n-2) \cdots (n-k+1) = \frac{n!}{(n-k)!} \qquad \text{(A-10)}$$

In particular, the number of permutations of n objects taken n at a time is

$$P(n,n) = n! \qquad \text{(A-10a)}$$

There should be no confusion between $P(n,k)$ and $P(n)$ because the former is the permutation of n objects taken k at a time and the latter is the probability of event n.

A combination of n different objects taken k at a time is a *selection* of k out of the n objects with no attention given to the order of arrangement. The number of combinations of n objects taken k at a time is denoted by $_nC_k$, $C_k{}^n$, $C_{n,k}$, $\binom{n}{k}$, or $C(n,k)$ and is given by

$$C(n,k) = \frac{n(n-1)\cdots(n-k+1)}{k!} = \frac{n!}{k!(n-k)!} = \frac{P(n,k)}{k!} \quad \text{(A-11)}$$

An alternative method for generating the binomial coefficients given by Eq. (A-11) is use of Pascal's triangle, Fig. A-1. Examination of this triangle shows that any number, with the exception of the ones, is generated by summing the adjoining two numbers in the line above it and between which it appears. This is indicated by lines in Fig. A-1.

```
n = 0                         1
n = 1                       1   1
n = 2                     1   2   1
n = 3                   1→3   3   1
n = 4                 1   4   6   4   1
n = 5               1   5  10  10→5   1
n = 6             1   6  15  20  15   6   1
n = 7           1   7  21  35→35  21   7   1
n = 8         1   8  28  56  70  56  28   8   1
```

FIGURE A-1
Pascal's triangle.

EXAMPLE A-8 A family consisting of a father, a mother, a daughter, and two sons attends a concert and sits in adjacent seats. (*a*) How many distinct seating arrangements exist? (*b*) How many arrangements if order is not considered?
(*a*) $P(5,5) = 5! = 120$
(*b*) Only one. No matter what the seating order, the family as a unit must be selected and there is only one way of doing that. ////

EXAMPLE A-9 How many different ways can a hand of 13 cards be selected from a standard bridge deck?

$$C(52,13) = \frac{52!}{13!39!} = 635{,}013{,}559{,}600 \qquad ////$$

A-7 RANDOM VARIABLES

A *random variable* is some *function* (real and single-valued) of the outcome of a "repetitive operation" in which we have a particular interest. If a pair of dice is thrown, the sum of the face values, the square of the sum of the face values, the sum of the squares of the face values, the product of the face values, the difference of the face values, the square of the difference of the face values, etc., are all random variables. If a sample of 100 light bulbs is drawn "at random" from a lot of 10,000, and tested, the number of defective bulbs in the sample is a random variable. If one bulb is taken at random from a large lot and burned, the time required for it to "expire" is a random variable.

More generally, we are dealing (in our observations or trials or experiments) with a quantity which, in some manner, characterizes the result of the operation performed. Each of these quantities can assume, no matter how homogeneous we try to make them, various values for various operations, which depend on random differences in the circumstances of these operations, these circumstances being beyond our control. In probability theory, such quantities are called *random* (or *stochastic*) *variables*.

Suppose we have a finite sample space S listing n elements $e_1, e_2, \ldots, e_n$. There are 2^n possible events which can be formed from these elements, provided we count the empty event ϕ and the entire space S as two of the events. This is clear from the fact that we have the choice of selecting or not selecting each of the n elements in making up an event. We rarely are interested in all 2^n events and their probabilities, but instead, we are usually interested in a relatively small number of them produced by specified values of some function defined on the elements.

A real and single-valued function $x(e)$ defined on each element e in the sample space is called a *random variable*. Assume $x(e)$ can take on only the values $x_1, x_2, \ldots, x_k$, and $E_1, E_2, \ldots, E_k$ are the events for which $x(e) = x_1, \ldots, x(e) = x_k$, respectively. Let $P(E_1) = p(x_1), \ldots, P(E_K) = p(x_k)$. Then we say that the random variable $x(e)$ is a *discrete random variable* which takes the values $x_1, \ldots, x_k$ with probabilities $p(x_1), \ldots, p(x_k)$, respectively.

Many sample spaces contain an infinite number of elements which cannot be put into one-to-one correspondence with probabilities. Such sample spaces have the property of a continuum. The random variables associated with such spaces assume values that represent continua and are known as *continuous random variables*. Under these circumstances, we are not able to associate probabilities with the different values which the random variable can assume in the same way as we can for a discrete random variable.

EXAMPLE A-10   Let x be a random variable giving the sum of the upper faces when two dice are thrown. Construct the probability function of x.

There are 36 different ways the two dice can be thrown. Each element in this sample space is equally likely, i.e., each has a probability of $\frac{1}{36}$. The probability function is

x	2	3	4	5	6	7	8	9	10	11	12
$p(x)$	$\frac{1}{36}$	$\frac{2}{36}$	$\frac{3}{36}$	$\frac{4}{36}$	$\frac{5}{36}$	$\frac{6}{36}$	$\frac{5}{36}$	$\frac{4}{36}$	$\frac{3}{36}$	$\frac{2}{36}$	$\frac{1}{36}$

////

A-8 DISTRIBUTION FUNCTIONS

The distribution function or probability law of a random variable provides the probability that the random variable has any specified value (or is less than any specified number) when the basic "repetitive operation" is performed once.

For the case of the discrete random variable X, the space of X consists of at most a countably infinite number of values, $x_1, x_2, \ldots$. With each possible outcome x_i we associate a number $p(x_i) = P(X = x_i)$, called the probability of x_i. The numbers $p(x_i)$, $i = 1, 2, \ldots$ must satisfy the following conditions:

$$\left. \begin{array}{l} (a)\ p(x_i) \geq 0 \quad \text{for all } i \\ (b)\ \sum_{i=1}^{\infty} p(x_i) = 1 \end{array} \right\} \quad \text{(A-12)}$$

The function p is called the *probability function* (or point probability function) of the random variable X. The collection of pairs $[x_i, p(x_i)]$, $i = 1, 2, \ldots$ is sometimes called the *probability distribution* of X.

For the case of the continuous random variable, the function p, defined only for the discrete values $x_1, x_2, \ldots$ is replaced by a function f defined for *all* values of x within the range (R_x) which is an interval or a collection of intervals. (which may be from $-\infty$ to $+\infty$). In this case, the *probability density function* f, denoted by pdf, is a function satisfying the following conditions:

$$(a)\ f(x) \geq 0 \quad \text{for all } x \text{ within } R_x \quad \text{(A-13)}$$
$$(b)\ \int_{R_x} f(x)\,dx = 1$$

Further consider the case of the random variable X which may be either discrete or continuous. We define F to be the *cumulative distribution function*, denoted by cdf, of the random variable X where $F(x) = P(X \leq x)$. If X is a discrete random variable,

$$F(x) = \sum_i p(x_i) \quad \text{(A-14)}$$

where the sum is taken over all indices i satisfying $x_i = x$. If X is a continuous random variable with pdf f,

$$F(x) = \int_{-\infty}^{x} f(x')\,dx' \quad \text{(A-15)}$$

The cdf has two important properties: (1) The function F is nondecreasing. That is, if $x_1 = x_2$, then $F(x_1) = F(x_2)$. (2) $\lim_{x \to -\infty} F(x) = 0$ and $\lim_{x \to \infty} F(x) = 1$. (This is often written as $F(-\infty) = 0$ and $F(\infty) = 1$.)

If we know the cdf of a random variable X, we can obtain its pdf (in the continuous case) and the point probabilities (in the discrete case) as follows: (1) Let F be the cdf of a continuous random variable with pdf f. Then

$$f(x) = \frac{d}{dx} F(x) \qquad \text{(A-16)}$$

for all x at which F is differentiable. (2) Let X be a discrete random variable with possible values $x_1, x_2, \ldots$, with $x_1 < x_2 < \cdots$. Let F be the cdf of X. Then

$$p(x_i) = P(X = x_i) = F(x_i) - F(x_{i-1}) \qquad \text{(A-17)}$$

EXAMPLE A-11 A continuous random variable has a cdf given by

$$F(x) = 0, \; x \leq 0$$
$$F(x) = 1 - e^{-x}, \; x > 0$$

Find the pdf.

$$\frac{dF(x)}{dx} = f(x) = e^{-x}, \; x \geq 0$$

$$= 0, \text{ elsewhere} \qquad ////$$

A-9 PARAMETERS OF A DISTRIBUTION

To accurately characterize a distribution, we must know the entire point probability function or the probability density function of the random variable. In most practical applications, the probability mass will be concentrated within a relatively narrow interval. Accordingly, to obtain some idea of the whole distribution, it is often appropriate to indicate the position of this interval by some measure of location by giving a typical value of x. Although any such measure is, as a rule, uniquely determined by the distribution function, the converse is (obviously) not true.

While there are many such measures of location, three are the most used: mode, median, and mean. The *mode* is defined as the most probable value, i.e., that value, class, or category which occurs the most often, i.e., with the highest frequency. In some sense, this represents the most typical. The *median* is defined as the mid-value, i.e., the value of the random variable for which the probability of finding a smaller value than the median is equal to the probability of finding a greater value. Thus the median is the value of the middle item (or the mean of the values of the two middle items) when the items are arranged in an increasing or decreasing order. The *mean* is defined as the center of gravity of the whole probability mass. This mean value is thus the arithmetic mean.

Which measure of location, or central tendency, to use is arbitrary and a matter of convenience. The mean, however, is the most commonly used.

The mean, as a measure of central tendency, is also known as the expectation, $E(X)$. By definition

$$E(X) \equiv \begin{cases} \sum_i x_i p(x_i) & \text{for discrete} \quad \text{(A-18)} \\ \int_{-\infty}^{\infty} x f(x)\, dx & \text{for continuous} \quad \text{(A-19)} \end{cases}$$

provided the sum or integral converges, i.e.,

$$\sum_i |x_i| p(x_i) < \infty \quad \text{or} \quad \int_{-\infty}^{\infty} |x| f(x)\, dx < \infty$$

otherwise undefined.

Expectation, $E(X)$, is a fixed number, i.e., a constant, which is often denoted by μ. If the frequency is interpreted as a mass distribution of total mass 1, then $E(X)$ represents the center of gravity of that mass distribution. Alternatively, $E(X)$ can be interpreted as the *weighted average* of the values of X, using the density function as the weighting function.

A distribution is not characterized by its location since every distribution will have a location, central tendency, or expectation. A further help in characterization (not sufficient although very useful in practice) is a measure of dispersion, i.e., how much the probability mass is spread about the chosen measure of central tendency. It should be emphasized that the distribution is not uniquely determined by the two parameters (measure of central tendency and measure of dispersion) alone. There are a large number of ways of determining a measure of dispersion but the most commonly used is variance V (which is the square of the standard deviation σ). By definition

$$\sigma^2 \equiv V(X) \quad \text{(A-20)}$$

$$V(X) = E[(X - \mu)^2] \quad \text{(A-21)}$$

$$V(X) = \begin{cases} \sum_i (x_i - \mu)^2 p(x_i) & \text{for discrete} \quad \text{(A-22)} \\ \int_{-\infty}^{\infty} (x - \mu)^2 f(x)\, dx & \text{for continuous} \quad \text{(A-23)} \end{cases}$$

As an alternative, computation of the variance can frequently be simplified as follows:

$$\begin{aligned} V(X) &= E[(X - \mu)^2] \\ &= E[X^2] - \mu^2 \\ &= E[X^2] - \{E[x]\}^2 \quad \text{(A-24)} \end{aligned}$$

It should be clearly understood that variance is a constant. It provides a measure of the spread of the distribution about its mean μ. If the variance is large, the distribution tends to be diffuse; if the variance is small, the distribution tends to be concentrated near the mean. In the same sense that the mean is determined from a calculation of the first moment of the distribution about the origin, the variance is analogous to the second moment of inertia taken about the center of gravity.

A-10 PROBABILITY DISTRIBUTIONS

There are at least 20 different point probability functions or probability density functions. Several of these have rather specific applicability in reliability theory and application.

A-10-1 Discrete Distributions (point probability functions)

A-10-1a Binomial distribution X is a discrete random variable with range $0, 1, 2, \ldots, n$ (n being a fixed integer)

$$p(x_i) = \Pr(X = x_i) = \binom{n}{i} p^i q^{n-i} \qquad \text{(A-25)}$$

for $i = 1, 2, \ldots, n$; where $p + q = 1$: $p(x_i) = 0$, otherwise

$$\mu = np \qquad \sigma^2 = np(1 - p) \qquad \text{(A-26)}$$

A-10-1b Poisson distribution X is a discrete random variable with range $0, 1, 2, \ldots$, such that

$$p(x_i) = \Pr(X = x_i) = \frac{\lambda^i e^{-\lambda}}{i!}; \qquad i = 1, 2, \ldots, \qquad \text{(A-27)}$$

$p(x_i) = 0$, otherwise;
  λ is a given positive parametric constant

$$\mu = \lambda \qquad \sigma^2 = \lambda \qquad \text{(A-28)}$$

The Poisson distribution is an approximation to the binomial distribution where $\lambda = np$ for large n and small p.

A-10-2 Continuous Distributions (probability density functions)

A-10-2a Gamma distribution

$$f(x) = \frac{a^n x^{n-1} e^{-ax}}{\Gamma(n)} \qquad \begin{array}{l} a > 0 \\ n > 0 \\ 0 \leq x \leq \infty \end{array} \qquad \text{(A-29)}$$

$$\left. \begin{array}{l} \mu = \dfrac{n}{a} \\[1em] \sigma^2 = \dfrac{n}{a^2} \end{array} \right\} \qquad \text{(A-30)}$$

A-10-2b Erlang distribution This is a special case of the gamma distribution where n is an integer.

A-10-2c Exponential (negative) distribution This is a special case of the gamma distribution where $n = 1$.

$$f(x) = ae^{-ax}$$
$$F(x) = 1 - e^{-ax} \qquad 0 \le x \le \infty \qquad \text{(A-31)}$$
$$\mu = 1/a \qquad \sigma^2 = 1/a^2 \qquad \text{(A-32)}$$

A-10-2d Normal (gaussian) distribution

$$f(x) = \frac{1}{\sigma \sqrt{2\pi}} e^{-[(x-\mu)^2/2\sigma^2]} \qquad -\infty < x < \infty \qquad \text{(A-33)}$$

where μ and σ are constants which are the mean and standard deviations of X, respectively. The cdf

$$F(x) = \int_{-\infty}^{x} f(x) \, dx$$

and the pdf are expressed in tabular form in a great number of mathematical tables. (The cdf is a nonelementary integral.)

Any normal distribution can be converted to the standard form by using a new variable

$$z = \frac{x - \mu}{\sigma} \qquad f(z) = \frac{1}{\sqrt{2\pi}} e^{(-z^2/2)} \qquad \text{(A-34)}$$

This standard normal distribution has a mean of 0 and a standard deviation of 1.

The pdf is symmetrical about the mean, with a maximum at the mean and inflection points at one standard deviation from the mean.

A-10-2e Lognormal distribution The lognormal distribution is closely related to the normal distribution. Assume X is normally distributed with mean μ and variance σ^2. Let $Y = e^X$. Then Y has the lognormal distribution, i.e., Y is lognormal if and only if $\ln Y$ is normal.

$$f(x) = \frac{1}{\beta \sqrt{2\pi}} x^{-1} \exp\left[-\frac{(\ln x - \alpha)^2}{2\beta^2}\right] \qquad \begin{matrix} x > 0 \\ \beta > 0 \end{matrix} \qquad \text{(A-35)}$$

$$f(x) = 0, \text{ elsewhere}$$
$$\mu = e^{[\alpha + (\beta^2/2)]}$$
$$\sigma^2 = e^{(2\alpha + \beta^2)}(e^{\beta^2} - 1) \qquad \text{(A-36)}$$

A-10-2f Weibull distribution This distribution was proposed in the late 1940s by W. Weibull. It has since been shown to apply to a large number of situations which are rather diverse. It has found extensive use in general engineering practice.

$$f(x) = \left(\frac{b}{\theta - x_0}\right)\left(\frac{x - x_0}{\theta - x_0}\right)^{b-1} \{e^{-[(x-x_0)/(\theta-x_0)]^b}\} \quad \text{(A-37)}$$

$$F(x) = 1 - e^{-[(x-x_0)/(\theta-x_0)]^b}$$

$$x > x_0$$
$$x_0 \geq 0$$
$$\theta > 0$$
$$b > 0$$

The three parameters (x_0, θ, b) of the Weibull distribution are usually determined experimentally from data obtained during the course of testing. They have physical significance:

x_0 = lower bound of variable x
θ = characteristic value, related to the median, thus is an indicator of central tendency
b = Weibull slope—a direct function of the uniformity of the item; the larger the slope, the greater the uniformity

A plot of $\ln \ln (1/[1 - F(x)])$ versus $\ln x$ is a straight line with slope b.

$$\mu = x_0 + (\theta - x_0)\Gamma(1 + 1/b) \quad \text{(A-38)}$$

$$\sigma^2 = (\theta - x_0)^2 \left[\Gamma\left(1 + \frac{2}{b}\right) - \Gamma\left(1 + \frac{1}{b}\right)\right]$$

When the Weibull slope is 1, this is a special case, i.e., the negative exponential distribution. When the Weibull slope is about 3.5, the Weibull distribution is a good approximation to the normal distribution.

PROBLEMS

A-1 Assume ten coins are tossed. What is the probability of getting at least one head? Of at least one tail?

A-2 An office has 100 calculating machines. Some of these are electric (E) while others are manual (M). Some are new (N) and others are used (U). The table gives the number in each category.

	E	M	Total
N	45	20	65
U	20	15	35
Total	65	35	100

An individual enters the office, picks a machine at random and finds that it is new. What is the probability that it is electric?

A-3 A president is to be elected from the membership of a political organization which has 100 members. If the ratio of male to female is 4:1 and half of both men and women are married, what is the probability that (*a*) the president is a man, (*b*) the president is a married woman, and (*c*) the president is a married man or married woman?

A-4 Events E_1, E_2, E_3, and E_4 are mutually exclusive outcomes of a random experiment. Are the following sets of probability values properly assigned?
(*a*) $P(E_1) = \frac{1}{6}$; $P(E_2) = \frac{1}{3}$; $P(E_3) = \frac{1}{2}$; $P(E_4) = 0$
(*b*) $P(E_1) = \frac{1}{4}$; $P(E_2) = -\frac{1}{4}$; $P(E_3) = P(E_4) = \frac{1}{2}$
(*c*) $P(E_1) = 1.6$; $P(E_2) = -.6$; $P(E_3) = P(E_4) = 0$

A-5 If $C(n,8) = C(n,12)$, find n.

A-6 How many permutations are possible with the letters in the words (*a*) copies, (*b*) collections, (*c*) assemblies, and (*d*) addressed, taken all together?

A-7 A box of 1000 rivets contains
    60 rivets with type A defects (and others)
    35 rivets with type B defects (and others)
    15 rivets with type C defects (and others)
    8 rivets with type A and type B defects
    6 rivets with type A and type C defects
    3 rivets with type B and type C defects
    2 rivets with all three types of defects
Select a rivet at random from the box.
(*a*) What is the probability that the rivet will have a type A defect or type B defect, or both?
(*b*) What is the probability that the rivet will have at least one of the three types of defect?
(*c*) What is the probability that the rivet will be free of any defect?

A-8 A rivet drawn from the box in Prob. A-7 is known to have a type B defect. What is the probability it also has a type C defect?

A-9 Three light bulbs in a box of eight have broken filaments. If the bulbs are tested at random, one at a time, what is the probability that the second defective bulb is found as soon as the fourth bulb is tested?

A-10 Employees in a factory work during three shifts. It has been observed that the first shift produces 40 percent, the second shift 30 percent, and the third shift 30 percent of the production. The number of defective items, however, are in the proportion 1:2:3, respectively by shifts. If articles produced by the three shifts are placed in one well-mixed pile and one item selected at random proves defective, what is the probability that the defective item was produced by (*a*) the first shift? (*b*) the second shift? (*c*) the third shift?

A-11 In examining a patient, a physician suspected that he has one of three illnesses: A_1, A_2, A_3. The probability of each, under the conditions is:

$$P_1 = 0.6, P_2 = 0.1, P_3 = 0.3$$

In order to make diagnosis more precise, some analysis is specified which yields a positive result with probability 0.2 in the case of A_1, probability 0.1 in the case of A_2, and probability 0.8 in the case of A_3. The analysis was carried out five times; it yielded a positive result four times and a negative result once. After the analysis has been made, what is the probability of each of the illnesses?

A-12 You are chairman of a committee to raise money for student activities and it is decided to do this through a dinner party of some sort. On the basis of past experience, it is expected that an outdoor picnic would yield a profit of $700 if it does not rain whereas an indoor buffet would yield $200. Alternatively, if it does rain, the outdoor picnic would yield $90 whereas the indoor buffet would yield $500. Which way should the dinner be given if the probability of rain on the scheduled day is (*a*) 0.2, (*b*) 0.6.

A-13 What is the probability of an atom (radioactive) being "alive" at time $t + t'$ provided it was known to be "alive" at time t? What conclusions can be drawn from this result, especially related to known biological processes?

A-14 The probability of the closing of each relay of the circuit shown is given by p. If all relays function independently, what is the probability that a current exists between the terminals L and R?

A-15 Consider a shaft and its bearing. The od of the shaft is 0.550-in. average, with a standard deviation of 0.0016 in. The id of the bearing is 0.559 in. average with a standard deviation of 0.0018 in. What is the probability of interference?

A-16 The life of a given part is known to have a negative exponential distribution with a mean life of 1000 h. What is the probability that a part will last less than 400 h? Less than 200 h?

A-17 Sheets of tin plate are first cut into strips approximately 5 in. wide, and these, in turn, are cut into sections which are rolled up and soldered together to make cylinders. The sections are rolled to make the 5-in. side the height of the cylinder. Numerous measurements show an actual height of 5.007 in. with a standard deviation of 0.009 in. The average diameter is 2.998 in. with a standard deviation of 0.020 in. Suppose it is not practical to measure the volume directly, but rather to estimate it from linear measurements. If the cylinders are capped to make cans, what fraction of the cans will have volumes in excess of 35 in^3? What fraction will have volumes less than 32 in^3?

A-18 A fair coin is tossed until a tail appears for the first time. (*a*) Find the probability function representing this random experiment. (*b*) Find the distribution function corresponding to the probability function and evaluate for $x = 5$.

A-19 A fair coin is tossed 10 times. (*a*) What is the probability of getting three heads? (*b*) What is the probability of getting more than three heads? (*c*) What difference would occur in the answers if 10 coins were tossed at once rather than one coin 10 times?

A-20 A regular and "unloaded" die is thrown five times. What is the probability of getting exactly *one ace*?

A-21 In a given manufacturing process 10 percent of all items turn out defective. If, from a day's output, three items are selected at random, what is the probability that all three would be defective?

A-22 The number of typographical errors per page prior to proofreading is assumed to obey the Poisson probability function with $k = 3$. Find the probability of finding at most one error, if a page is examined at random.

A-23 The life of a certain make of light bulb obeys the exponential probability function. If the mean life is claimed to be 900 h, what is the probability of the lamp lasting (*a*) over 900 h? (*b*) over 1100 h?

A-24 Given $f(x) = x - x^3$, $0 \leq x \leq 1$
  $f(x) = 0$ otherwise
(*a*) Graph the function.
(*b*) Is the function a pdf?
(*c*) If not, what must be done to make it a pdf?

A-25 Given $f(x) = x/3$ $0 < x < 1$
  $f(x) = \frac{1}{3}$ $1 < x < 3$
  $f(x) = (4 - x)/3$ $3 < x < 4$
  $f(x) = 0$ otherwise
(*a*) Graph the function.
(*b*) Is the function a pdf?
(*c*) If not, what must be done to make it a pdf?

A-26 Given $f(x) = (\frac{5}{2})(x - 1)^4$ $0 < x < 2$
  $f(x) = 0$ otherwise
(*a*) Graph the function.

(b) Is the function a pdf? (c) Make the function a pdf, if it is not now a pdf.
(d) Find the cdf. (e) Graph the cdf. (f) Find $P(0.5 < x < 1)$.

A-27 A simplified slot machine has two dials. Each dial has three pictures on it, identified as peaches, pears, and cherries. The machine operates so that the two dials operate independently. After they are spun, each comes to rest with one of the three pictures showing in a window on the front of the machine. The probabilities of possible outcomes, for each dial, are

Outcomes	Peaches	Pears	Cherries
Probabilities	0.60	0.30	0.10

It costs 10 cents for each play, consisting of pulling a lever that spins the dials, resulting in one of nine possible combinations of two pictures, one on each dial. The machine pays off thus:

For 2 peaches	10 cents	For 2 pears	30 cents
For 2 cherries	$1.00	No payoff for any other combination	

Find the expectation of net profit (in money) for a person who plays once.

A-28 Let X be the life (in hours) of a certain type of light bulb. Assuming X is a continuous random variable, the pdf of X is given by

$$f(x) = a/x^3 \quad 1000 \leq x \leq 2000 \quad f(x) = 0 \quad \text{elsewhere}$$

If the foregoing function is not a true pdf, make it one, if possible. Sketch the function and the cdf. What is the equation for the cdf?

BIBLIOGRAPHY

References which do not use set theory

A-1 ALDER, H. L., and E. B. ROESSLER: "Probability and Statistics," 4th ed., W. H. Freeman and Company, San Francisco, 1968.

A-2 ARLEY, N., and K. R. BUCH: "Theory of Probability and Statistics," John Wiley & Sons, Inc. New York, 1950.

A-3 GNEDENKO, B. V., and A. YA. KINCHIN: "Theory of Probability," Dover Publications, Inc., New York, 1962.

A-4 MOSTELLER, F., R. E. K. ROURKE, and G. B. THOMAS: "Probability with Statistical Applications," Addison-Wesley Publishing Company, Inc., Reading, Mass., 1961.

A-5 SPIEGEL, M. R.: "Statistics," McGraw-Hill Book Company, New York, 1961.

References which use set theory

A-6 DRAKE, A. W.: "Fundamentals of Applied Probability Theory," McGraw-Hill Book Company, New York, 1967.

A-7 DUBES, R. C.: "Theory of Applied Probability," Prentice-Hall, Inc., Englewood Cliffs, N.J., 1968.

A-8 GRAY, H. L., and P. L. ODELL: "Probability for Practicing Engineers," Barnes & Noble, Inc., New York, 1970.

A-9 GUTTMAN, I., and S. S. WILKS: "Introductory Engineering Statistics," John Wiley & Sons, Inc., New York, 1965.

A-10 LIPSCHUTZ, S.: "Probability," McGraw-Hill Book Company, New York, 1968.

A-11 MAKSOUDIAN, Y. L.: "Probability and Statistics with Applications," International Textbook Company, Scranton, Pa., 1969.

A-12 MEYER, P. L.: "Introductory Probability—Statistical Applications," Addison-Wesley Publishing Company, Inc., Reading, Mass., 1965.

A-13 MILLER, I., and J. E. FREUND: "Probability and Statistics for Engineers," Prentice-Hall, Inc., Englewood Cliffs, N.J., 1965.

APPENDIX B

EQUIPMENT AND FACTORS IN RELIABILITY TESTING

B-1 GENERALITIES

A Test Equipment Includes

 1 Equipment for simulating environment.
 2 Instruments for measuring and controlling environment.
 3 Instruments for measuring sample's performance.
 4 Fixtures, cables, etc., for connecting sample to test equipment.

B Factors to Consider

 1 Effects of environment on typical test samples.
 2 Means of simulation.
 3 Measurement and control instrumentation.
 4 Other considerations.

C Environment Categories

 1 Natural or induced (i.e., unnatural). Latter induced by operation of sample or the system that includes it.
 2 Atmospheric or nonatmospheric.

B-2 WHY SIMULATE REALITY?

1 Local test facilities cost money but save travel expenses.
2 Flight tests are very expensive and may be destructive (purposely or not).
3 Telemetry and on-board recorders are limited in capacity and are more unreliable than lab equipment.
4 Testing can minimize hazards and prevent loss of life.
5 Environment can be programmed as required.
6 Effects of single environments can be studied.

B-3 TEMPERATURE ENVIRONMENTS

A High Temperature

1 Effects
  (*a*) Expedites chemical reactions.
  (*b*) Temperature coefficients.
  (*c*) Weakens structures.

2 Simulation—Insulated chamber, heater (electrical), fan.

3 Measurement and control
  (*a*) Thermocouples, gas-bulb thermometers or thermistors.
  (*b*) Exposed to airflow.
  (*c*) Manual controls, strip-chart recorder-controller, or cam-programmed circular-chart controller.   Latter preferred because it cuts labor costs.
  (*d*) Fail-safe control essential.

4 Other considerations
  (*a*) Design electrical test cables to withstand temperature.
  (*b*) Allow temperature to stabilize before performance tests.
  (*c*) Do not operate sample longer than necessary to measure its performance (heating).

B Low Temperature

1 Effects
  (*a*) Retards chemical reactions (battery problems).
  (*b*) Temperature coefficients.
  (*c*) Embrittlement.
  (*d*) Loss of lubrication.

2 Simulation
  (*a*) Insulated chamber and fan (like high temperature).
  (*b*) Cooled by mechanical refrigeration, expanding liquid or solid gases, or (lately) thermoelectric cooling.

3 Measurement and control similar to high temperature.

4 Other considerations
   (*a*) Cable and stabilization problems like high temperature.
   (*b*) Frost forms when chamber door opened, and may be undesirable.

C Temperature Shock

1 Is usually several rapid (5 min or less) transitions between high- and low-temperature extremes. Sample always tested after exposure, not during exposure.

2 Effects
   (*a*) Structural failure from different temperatures within sample.
   (*b*) Temperature coefficients.

3 Simulation by transferring between high- and low-temperature chambers.

4 Measurement and control of chambers as described above.

5 Other considerations—Tested after exposures completed. Last exposure before room temperature checkout is low temperature, assuring frost.

D Sunshine

1 Effects
   (*a*) High-temperature effects induced on irradiated surface.
   (*b*) Ultraviolet radiation causes photo-degradation of materials.

2 Simulation—special lamps irradiate sample.

3 Measurement and control—maintain constant power.

4 Other considerations—sample tested after exposure.

B-4 PRESSURE ENVIRONMENTS

A Low Pressure

1 Effects
   (*a*) Fluids and gases leak through poor seals.
   (*b*) Rupture of pressurized containers.
   (*c*) Arcing or corona.
   (*d*) No cooling by convection.

2 Simulation—chamber and vacuum pump, perhaps diffusion pump.

3 Measurement and control
   (*a*) McLeod, Pirani, Redhead, etc., gauges.
   (*b*) Barometers.
   (*c*) Strain-gauge transducers with strip or circular chart controllers.

4 Other considerations
  (*a*) Shafts (thrust and rotary) and cables must be sealed.
  (*b*) Do not operate the sample longer than necessary to measure its performance (heating).

B Decompression

1 Is pertinent only to equipment that must perform during or after rapid changes in altitudes.
2 Effects—rupture of sealed containers.
3 Simulation—vacuum or diffusion pumps are not fast enough. Couple small test chamber to large evacuated low-pressure chamber, open valve between quickly.
4 Measurement and control—compute from volumes and pressures. Measure control pressure as in low-pressure chambers.
5 Other considerations—if using glass bell jar, put protective covering over it because it might fail!

B-5 CORROSIVE/EROSIVE ENVIRONMENTS

A Rain

1 Effects
  (*a*) Low resistance between electrical circuits.
  (*b*) Rust and other corrosion.
  (*c*) Erosion if vehicles fly through it at high speed.

2 Simulation
  (*a*) Pressurized water flowing through sprinkler nozzles (uniform distribution difficult).
  (*b*) Tank with capillary tubes uniformly distributed over roof of chamber.
  (*c*) Rainfall should be 45° from vertical per Mil-STD-810A.

3 Measurement and control
  (*a*) Precalibrate, measuring rate with rainfall gauges and noting pressure and flow rate at input.
  (*b*) Temperature and droplet size must be controlled.

4 Special considerations
  (*a*) Create over sled or centrifuge to determine in-flight corrosion.
  (*b*) Plastic bags over connectors that are usually protected. Dummy plugs in receptacles.
  (*c*) Tested after exposure.

B Humidity

1 Effects
  (*a*) Rust and other corrosion.
  (*b*) Low resistance between electrical circuits.
  (*c*) Absorption of moisture with resultant swelling.

2 Simulation
  (*a*) Closed chamber, fan to circulate humid air.
  (*b*) Moisture added as steam, as spray of water (not on sample), or evaporated from heated pan. Distilled water required.
  (*c*) Is generally an accelerated, 10-day cyclic type environment in which temperature is raised, then lowered, then raised, etc., to cause condensation of dew on sample.

3 Measurement and control—Wet and dry thermocouples together with two conventional strip or circular chart recorder controllers. Somewhat simplified by diffcrential sensor that detects difference between wet and dry temperatures, used in conjunction with dry thermocouple.

4 Special considerations
  (*a*) Test cables must be moisture-resistant.
  (*b*) Sample may be tested during and/or after exposure.

C Fungi

1 Effects—fungi secrete enzymes which can destroy many organic substances, including leather, paper, and lens coatings, as well as minerals.

2 Simulation is conducted in a heated, humid chamber. A spore suspension and nutrient material are sprayed on the sample. Conditions are maintained for a period of 14 to 28 days. The sample is then withdrawn and tested.

3 Measurement and control—conventional temperature and humidity instruments. Nothing for the fungi solutions except a pH meter, since standard solutions are readily available.

4 Special considerations
  (*a*) A formal statement that the sample contains no nutrient materials often obviates this test.
  (*b*) The test makes a mess in the chamber that is difficult to remove. The test is often subcontracted to commercial testing laboratories.
  (*c*) Test sample after exposure.

D Salt Fog

1 Effects include corrosion and binding of mechanisms (because of deposits).

2 Simulation is accomplished by spraying an atomized salt water solution into a heated, humid chamber. This is an accelerated test that lasts 48 hours or more.

3 Measurement and control consists of conventional temperature and humidity instrumentation plus apparatus for measuring pH and salt concentration.

4 Special considerations
  (*a*) This test is at best an arbitrary, controversial standard because it bears little resemblance to reality.
  (*b*) The chamber is highly specialized and is messy. It cannot be used easily for other types of tests, so salt fog tests are often subcontracted to commercial testing laboratories.
  (*c*) Test sample after exposure.

E **Sand and Dust**

1 Effects
  (*a*) Abrasion.
  (*b*) Binding in mechanisms.
  (*c*) Interference with electrical contacts.
  (*d*) Clogging of air filters.

2 Simulation is accomplished with a small, closed-cycle wind tunnel. Sand and dust particles of specific properties (readily available) are introduced into an airstream whose temperature, humidity, and speed are controlled.

3 Measurement and control
  (*a*) Conventional temperature and humidity controls.
  (*b*) Density of sand and dust controlled with photoelectric device; e.g., smoke gauge.
  (*c*) Airspeed measured with hot-wire anemometer.

4 Special considerations
  (*a*) An expensive, messy low-utilization facility; often subcontracted to a commercial testing laboratory.
  (*b*) Test sample after exposure.

B-6 MECHANICAL ENVIRONMENTS

A **Sustained Acceleration** (>0.1 second)

1 Effects include structural deformation, undesirable flow of fluids, and improper operation of spring-loaded devices; e.g., relays and check valves.

2 Simulation accomplished by centrifuge or by rocket-propelled sled on track.

3 Measurement and control in centrifuge by rpm and distance from center of rotation; in sled by thrust control and on-board accelerometers, or by doppler system.

4 Special considerations
   (*a*) Observing operation on centrifuge requires closed-circuit TV or complex optical system.
   (*b*) Department of Defense provides facilities for sled tests.
   (*c*) Acceleration gradient along radius of centrifuge must be considered.
   (*d*) Special centrifuges providing programmed acceleration are available. They use controllable clutches or devices that move sample along arm.
   (*e*) Test sample during and/or after exposure.

B   Vibration

1 Effects include structural weakening or failure, improper operation of spring-loaded devices (e.g., relays) and of other devices; e.g., electrical short circuits between adjacent conductors or vacuum tube elements.

2 Simulation methods vary considerably, depending on required displacement and frequency range.
   (*a*) Low-frequency, low-displacement vibration—motor drives table through eccentric counterweights or cams.
   (*b*) Low-frequency, high-displacement vibration—hydraulic piston controlled by electrically actuated valve.
   (*c*) High-frequency, low-displacement vibration—amplifier-driven loudspeaker device having table instead of cone.

3 Measurement and control by measuring displacement, velocity or acceleration, of spring-mounted seismic mass. There are many types of transducers.

4 Special considerations
   (*a*) Parameters are complex.
   (*b*) Equipment is expensive (except motor-driven type).
   (*c*) Is a harsh environment, so instrumentation should be calibrated often.
   (*d*) Random vibration commonly required, necessitating very expensive equipment that subdivides frequency range into many channels with filters.
   (*e*) Test sample during and/or after exposure.

C   Acoustic Field

1 Effects are similar to those of vibration and are induced in the sample by vibrating air particles.

2 Simulation by loudspeaker, by sirens, or by modulating a high-pressure, high-volume airstream with a controllable valve. Soundproof chamber required to protect hearing.

3 Measurement and control equipment similar to vibration equipment, with microphones substituted for accelerometers.

4 Special considerations
  (*a*) Expensive; may be subcontracted.
  (*b*) Required primarily for equipment in (or propelled by) rocket motors.
  (*c*) Supplements, does not replace, vibration tests.
  (*d*) Test sample during and/or after exposure.

D Shock

1 Effects include transient vibration, structural failure, momentary improper operation; e.g., electrical short circuits.

2 Simulation by accelerating, then decelerating at different rates:
  (*a*) Table, falling at 1 g, impacts on sand, spring, etc. It may fall vertically or may roll along on inclined ramp.
  (*b*) Table may be accelerated with a pneumatic or hydraulic piston and then braked (at low-gravity level) to a stop.
  (*c*) Sample may be merely dropped on concrete floor.
  (*d*) Waveshapes include half-sine wave, square wave, sawtooth, trapezoids.

3 Measurement and control—mechanical geometry (distance, pressure, materials) controlled. Measured with accelerometers. Ringing may be a problem

4 Special considerations
  (*a*) May be hazardous.
  (*b*) Test sample during and/or after exposure.

B-7 COMBINED ENVIRONMENTS

A Explosive Vapors

1 Affects safety in that explosion in aircraft or missile can be caused by motor brushes or by breaking electrical connections.

2 Simulation—pressure container filled with explosive vapors at various combinations of temperature and altitude.

3 Measurement and control—usual temperature and low pressure devices, plus explosive vapor detector.

4 Special considerations
  (*a*) Is a versatile facility in which small pyrotechnic devices can be tested.
  (*b*) Sample operated during exposure.

B Others

1 Vibration tests are often performed at high and low temperatures, as are shock tests.

2 Low pressure often combined with high and low temperature.

B-8 TEST SEQUENCE

1 Temperature, vibration, and humidity cause majority of failures.

2 Test exposure to severe environments can cause equipment to be unrealistically sensitive to moderate environments.

3 Test procedure sequence must thus be chosen with care.
    (*a*) Military specifications suggest an appropriate order.
        i Temperature and low pressure.
        ii Humidity, rain and other corrosive/erosive environments.
        iii Acceleration, vibration, shock, etc.

APPENDIX C
CONCEPT OF RANKING

The usual method of estimating populations from samples is to put the data into a frequency table and construct a frequency histogram which may be smoothed into a frequency curve. This is not reliable for small samples, since the results vary too much with a change in size of class interval and, generally, false peaks cannot be distinguished from true peaks. With small samples, it is better to order the data in increasing order of value and construct a cumulative plot. In other words, the estimation of the integral of the frequency function (cdf) is easier than estimation of the frequency function.

In order to make a cumulative plot, it is necessary to decide what range to assign to each failure in a group. The first failure in the set tested would have a definite fraction of the total population failing before it, if the entire population were tested. If we know exactly the percent of the population failing before the first failure in a set (e.g., 10 specimens), that percent would be the true rank of the first failure (in ten). Since we do not know the true rank, we must estimate it. We use an estimate such that, in the long run, the positive and negative errors of the estimate cancel each other. In other words, half the time the rank would be too high and the other half of the time the rank would be too low. A rank with this property is called a median rank. See Table C-1.

Table C-1 TABLE OF MEDIAN RANKS

Sample size = n

r	1	2	3	4	5	6	7	8	9	10
1	.5000	.2929	.2063	.1591	.1294	.1091	.0943	.0830	.0741	.0670
2		.7071	.5000	.3864	.3147	.2655	.2295	.2021	.1806	.1632
3			.7937	.6136	.5000	.4218	.3648	.3213	.2871	.2594
4				.8409	.6853	.5782	.5000	.4404	.3935	.3557
5					.8706	.7345	.6352	.5596	.5000	.4519
6						.8909	.7705	.6787	.6065	.5481
7							.9057	.7979	.7129	.6443
8								.9170	.8194	.7406
9									.9259	.8368
10										.9330

Sample size = n

r	11	12	13	14	15	16	17	18	19	20
1	.0611	.0561	.0519	.0483	.0452	.0424	.0400	.0378	.0358	.0341
2	.1489	.1368	.1266	.1178	.1101	.1034	.0975	.0922	.0874	.0831
3	.2366	.2175	.2013	.1873	.1751	.1644	.1550	.1465	.1390	.1322
4	.3244	.2982	.2760	.2568	.2401	.2254	.2125	.2009	.1905	.1812
5	.4122	.3789	.3506	.3263	.3051	.2865	.2700	.2553	.2421	.2302
6	.5000	.4596	.4253	.3958	.3700	.3475	.3275	.3097	.2937	.2793
7	.5878	.5404	.5000	.4653	.4350	.4085	.3850	.3641	.3453	.3283
8	.6756	.6211	.5747	.5347	.5000	.4695	.4425	.4184	.3968	.3774
9	.7634	.7018	.6494	.6042	.5650	.5305	.5000	.4728	.4484	.4264
10	.8511	.7825	.7240	.6737	.6300	.5915	.5575	.5272	.5000	.4755
11	.9389	.8632	.7987	.7432	.6949	.6525	.6150	.5816	.5516	.5245
12		.9439	.8734	.8127	.7599	.7135	.6725	.6359	.6032	.5736
13			.9481	.8822	.8249	.7746	.7300	.6903	.6547	.6226
14				.9517	.8899	.8356	.7875	.7447	.7063	.6717
15					.9548	.8966	.8450	.7991	.7579	.7207
16						.9576	.9025	.8535	.8095	.7698
17							.9600	.9078	.8610	.8188
18								.9622	.9126	.8678
19									.9642	.9169
20										.9659

Table C-1 TABLE OF MEDIAN RANKS (Continued)

Sample size = n

r	21	22	23	24	25	26	27	28	29	30
1	.0330	.0315	.0301	.0288	.0277	.0266	.0256	.0247	.0239	.0231
2	.0797	.0761	.0728	.0698	.0670	.0645	.0621	.0599	.0579	.0559
3	.1264	.1207	.1155	.1108	.1064	.1023	.0986	.0951	.0919	.0888
4	.1731	.1653	.1582	.1517	.1457	.1402	.1351	.1303	.1259	.1217
5	.2198	.2099	.2009	.1927	.1851	.1781	.1716	.1655	.1599	.1546
6	.2665	.2545	.2437	.2337	.2245	.2159	.2081	.2007	.1939	.1875
7	.3132	.2992	.2864	.2746	.2638	.2538	.2445	.2359	.2279	.2204
8	.3599	.3438	.3291	.3156	.3032	.2917	.2810	.2711	.2619	.2533
9	.4066	.3884	.3718	.3566	.3425	.3295	.3175	.3063	.2959	.2862
10	.4533	.4330	.4145	.3975	.3819	.3674	.3540	.3415	.3299	.3191
11	.5000	.4776	.4572	.4385	.4212	.4053	.3905	.3767	.3639	.3519
12	.5466	.5223	.5000	.4795	.4606	.4431	.4270	.4119	.3979	.3848
13	.5933	.5669	.5427	.5204	.5000	.4810	.4635	.4471	.4319	.4177
14	.6400	.6115	.5854	.5614	.5393	.5189	.5000	.4823	.4659	.4506
15	.6867	.6561	.6281	.6024	.5787	.5568	.5364	.5176	.5000	.4835
16	.7334	.7007	.6708	.6433	.6180	.5946	.5729	.5528	.5340	.5164
17	.7801	.7454	.7135	.6843	.6574	.6325	.6094	.5880	.5680	.5493
18	.8268	.7900	.7562	.7253	.6967	.6704	.6459	.6232	.6020	.5822
19	.8735	.8346	.7990	.7662	.7361	.7082	.6824	.6584	.6360	.6151
20	.9202	.8792	.8417	.8072	.7754	.7461	.7189	.6936	.6700	.6480
21	.9669	.9238	.8844	.8482	.8148	.7840	.7554	.7288	.7040	.6808
22		.9684	.9271	.8891	.8542	.8218	.7918	.7640	.7380	.7137
23			.9698	.9301	.8935	.8597	.8283	.7992	.7720	.7466
24				.9711	.9329	.8976	.8648	.8344	.8060	.7795
25					.9722	.9354	.9013	.8696	.8400	.8124
26						.9733	.9378	.9048	.8740	.8453
27							.9743	.9400	.9080	.8782
28								.9752	.9420	.9111
29									.9760	.9440
30										.9768

APPLICATION OF MEDIAN RANKS TO WEIBULL DISTRIBUTION

Assume 5 items are tested and fail (in hours) as indicated: 30, 18, 12, 25, 32.

Rank in increasing order	Histogram with smoothing	Median ranking	5% rank	95% rank
12	20	12.94	1.02	45.07
18	40	31.47	7.64	65.74
25	60	50.00	18.93	81.07
30	80	68.53	34.26	92.36
32	100	87.06	54.93	98.98

A plot of time to failure versus median ranking on Weibull paper will give a "straight" line (Fig. C-1). Once a plot is made, there is a question as to the

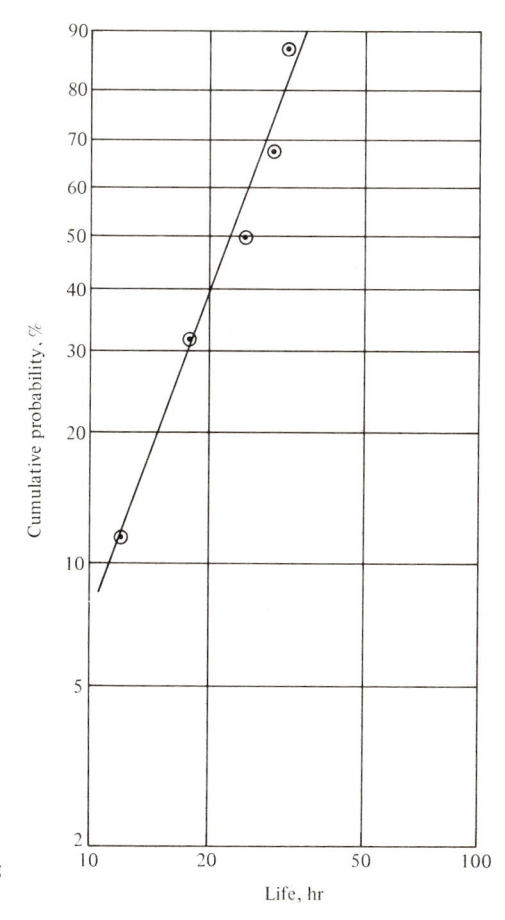

FIGURE C-1
Plot of data for median life showing application of median ranks.

240 INTRODUCTION TO RELIABILITY IN DESIGN

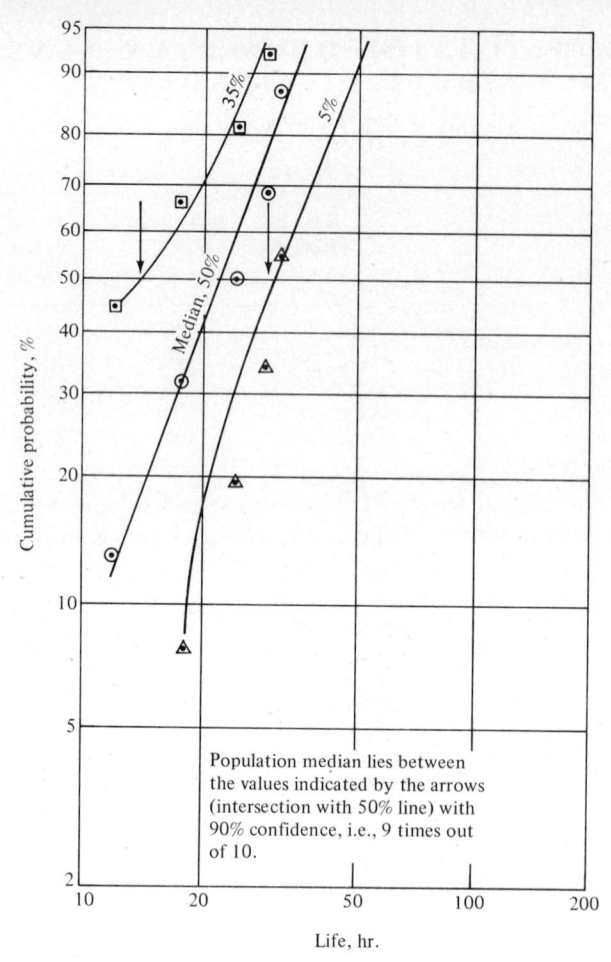

FIGURE C-2
Plot of data for median life and 90 percent confidence limits, showing application of 5 percent, median, and 95 percent ranks.

estimate of the population. The curve already plotted is the median life, 50 percent. What then is the range within which the true population median will lie, with some specific probability, e.g., 90 percent?

Confidence levels on median life, mean life, characteristic life, 10 percent life, etc., can be placed through use of other rank tables. Table C-2 gives 5 percent and 95 percent ranks and thus by use of this table we can obtain 90 percent confidence levels. Thus, if the median (50 percent) life of a given sample crosses the 95 percent rank at 0.6×10^5 cycles and the 5 percent rank at 2.8×10^5 cycles, we can say with 90 percent confidence that the median of the population lies somewhere between 0.6 and 2.8×10^5 cycles (Fig. C-2).

Table C-2a: TABLE OF 5 PERCENT RANKS

	Sample size = n									
	1	2	3	4	5	6	7	8	9	10
1	.0500	.0253	.0170	.0127	.0102	.0085	.0074	.0065	.0057	.0051
2		.2236	.1354	.0976	.0764	.0629	.0534	.0468	.0410	.0368
3			.3684	.2486	.1893	.1532	.1287	.1111	.0978	.0873
4				.4729	.3426	.2713	.2253	.1929	.1688	.1500
5					.5493	.4182	.3413	.2892	.2514	.2224
r 6						.6070	.4793	.4003	.3449	.3035
7							.6518	.5293	.4504	.3934
8								.6877	.5709	.4931
9									.7169	.6058
10										.7411

	Sample size = n									
	11	12	13	14	15	16	17	18	19	20
1	.0047	.0043	.0040	.0037	.0034	.0032	.0030	.0029	.0028	.0026
2	.0333	.0307	.0281	.0263	.0245	.0227	.0216	.0205	.0194	.0183
3	.0800	.0719	.0665	.0611	.0574	.0536	.0499	.0476	.0452	.0429
4	.1363	.1245	.1127	.1047	.0967	.0910	.0854	.0797	.0761	.0725
5	.2007	.1824	.1671	.1527	.1424	.1321	.1247	.1173	.1099	.1051
6	.2713	.2465	.2255	.2082	.1909	.1786	.1664	.1575	.1485	.1396
7	.3498	.3152	.2883	.2652	.2459	.2267	.2128	.1990	.1887	.1785
8	.4356	.3909	.3548	.3263	.3016	.2805	.2601	.2449	.2298	.2183
9	.5299	.4727	.4274	.3904	.3608	.3350	.3131	.2912	.2749	.2587
r 10	.6356	.5619	.5054	.4600	.4226	.3922	.3542	.3429	.3201	.3029
11	.7616	.6613	.5899	.5343	.4893	.4517	.4208	.3937	.3703	.3469
12		.7791	.6837	.6146	.5602	.5156	.4781	.4460	.4196	.3957
13			.7942	.7033	.6366	.5834	.5395	.5022	.4711	.4434
14				.8074	.7206	.6562	.6044	.5611	.5242	.4932
15					.8190	.7360	.6738	.6233	.5809	.5444
16						.8274	.7475	.6871	.6379	.5964
17							.8358	.7589	.7005	.6525
18								.8441	.7704	.7138
19									.8525	.7818
20										.8609

Table C-2b: **TABLE OF 95 PERCENT RANKS**

Sample size = n

	1	2	3	4	5	6	7	8	9	10
1	.9500	.7764	.6316	.5271	.4507	.3930	.3482	.3123	.2831	.2589
2		.9747	.8646	.7514	.6574	.5818	.5207	.4707	.4291	.3942
3			.9830	.9024	.8107	.7287	.6587	.5997	.5496	.5069
4				.9873	.9236	.8468	.7747	.7108	.6551	.6076
5					.9898	.9371	.8713	.8071	.7486	.6965
6						.9915	.9466	.8889	.8312	.7776
7							.9926	.9532	.9032	.8500
8								.9935	.9590	.9127
9									.9943	.9632
10										.9949

(r is the row index)

Sample size = n

	11	12	13	14	15	16	17	18	19	20
1	.2384	.2209	.2058	.1926	.1810	.1726	.1642	.1559	.1475	.1391
2	.3644	.3387	.3163	.2967	.2794	.2640	.2525	.2411	.2296	.2182
3	.4701	.4381	.4101	.3854	.3634	.3438	.3262	.3129	.2995	.2862
4	.5644	.5273	.4946	.4657	.4398	.4166	.3956	.3767	.3621	.3475
5	.6502	.6091	.5726	.5400	.5107	.4844	.4605	.4389	.4191	.4036
6	.7287	.6848	.6452	.6096	.5774	.5483	.5219	.4978	.4758	.4556
7	.7993	.7535	.7117	.6737	.6392	.6078	.5792	.5540	.5289	.5068
8	.8637	.8176	.7745	.7348	.6984	.6650	.6458	.6063	.5804	.5566
9	.9200	.8755	.8329	.7918	.7541	.7195	.6869	.6571	.6297	.6043
10	.9667	.9281	.8873	.8473	.8091	.7733	.7399	.7088	.6799	.6531
11	.9953	.9693	.9335	.8953	.8576	.8214	.7872	.7551	.7251	.6971
12		.9957	.9719	.9389	.9033	.8679	.8336	.8010	.7702	.7413
13			.9960	.9737	.9426	.9090	.8753	.8425	.8113	.7817
14				.9963	.9755	.9464	.9146	.8827	.8525	.8215
15					.9966	.9773	.9501	.9203	.8901	.8604
16						.9968	.9784	.9534	.9239	.8949
17							.9970	.9795	.9548	.9275
18								.9971	.9806	.9571
19									.9972	.9817
20										.9974

APPENDIX D
ALGEBRA OF NORMAL DISTRIBUTION FUNCTIONS†

BINARY OPERATIONS— INDEPENDENT DISTRIBUTIONS

Addition $(x + y)$

$$\mu_{x+y} = \mu_x + \mu_y$$
$$\sigma_{x+y} = (\sigma_x^2 + \sigma_y^2)^{1/2}$$

Subtraction $(x - y)$

$$\mu_{x-y} = \mu_x - \mu_y$$
$$\sigma_{x-y} = (\sigma_x^2 + \sigma_y^2)^{1/2}$$

Multiplication (xy)

$$\mu_{xy} = \mu_x \mu_y$$
$$\sigma_{xy} = (\mu_x^2 \sigma_y^2 + \mu_y^2 \sigma_x^2 + \sigma_x^2 \sigma_y^2)^{1/2}$$

Division (x/y)

$$\mu_{x/y} = \mu_x/\mu_y$$
$$\sigma_{x/y} = \frac{1}{\mu_y}\left(\frac{\mu_x^2 \sigma_y^2 + \mu_y^2 \sigma_x^2}{\mu_y^2 + \sigma_y^2}\right)^{1/2}$$

† See Reference (18-1).

Square (x^2)

$$\mu_{x^2} = \mu_x^2 + \sigma_x^2$$
$$\sigma_{x^2} = (4\mu_x^2\sigma_x^2 + 2\sigma_x^4)^{1/2}$$

Square root ($x^{1/2}$)

$$\mu_{x^{1/2}} = [\tfrac{1}{2}(4\mu_x^2 - 2\sigma_x^2)^{1/2}]^{1/2}$$
$$\sigma_{x^{1/2}} = [\mu_x - \tfrac{1}{2}(4\mu_x^2 - 2\sigma_x^2)^{1/2}]^{1/2}$$

Cube (x^3)

$$\mu_{x^3} = \mu_x^3 + 3\mu_x\sigma_x^2$$
$$\sigma_{x^3} = 3\mu_x^2\sigma_x$$

PROBLEMS

D-1 An electrical product requires four resistances in series. If the four resistances are as given below (with natural tolerances), what is the resistance of the series? What is the tolerance? What is the standard deviation?
20.0 ± 1.0; 19.0 ± 0.9; 23.0 ± 1.4; 24.0 ± 1.6 (ohms)

D-2 An assembly is composed of five components which are placed end to end. The lengths of the five components are 0.600, 0.510, 0.200, 0.800, and 0.310 in., respectively. The finished assembly can have a length of 2.420 ± 0.040 in. Assuming that all five components have a common variance, determine the specification limits on the individual components.

D-3 A shaft has a diameter of 1.148 ± 0.0040 in. The inside diameter of the mating bearing is 1.159 ± 0.0050 in. Assuming normal distributions, what is the minimum clearance that can be expected with random matching of shafts and bearings in assembly? What is the maximum expected clearance?

D-4 The transformer-amplifier arrangement shown has a specification on the output voltage of 75 ± 3 percent. Can the circuit meet the requirements?
$N = (2$ to $1) \pm 1$ percent $K = 3 \pm 2$ percent

Input voltage $E_i = 50 \pm 0.5$

Transformer N

Amplifier K

Output voltage E_o

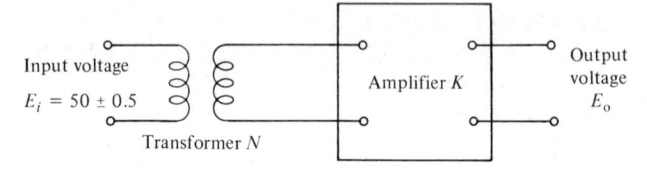

D-5 An emf of E volts drives a current of I amperes through a resistance of R ohms. According to Ohm's law, $E = IR$. If E and R are normally distributed with natural tolerances as follows
$E = 80 \pm 4$ V $R = 8 \pm 0.4\ \Omega$
find the natural tolerance of I.

D-6 The width of a slot on a duralumin forging is normally distributed with a mean of 0.8000 in. and a standard deviation of 0.0030 in. The specification limits are given as 0.8000 ± 0.0050 in.
(a) What percentage will be defective? What is the probability that a forging selected at random will be defective?
(b) What is the maximum standard deviation to yield not more than 1 part per 1000 defective?

D-7 The standard deviation in diameter is 0.001 in. for both the shaft and bearing in an assembly. The mean distance between centers of the shaft and bearing is 0.004 in. What is the probability of interference occurring during assembly?

D-8 Two resistors are assembled in series. They are rated at 15 Ω with a standard deviation of 0.5 Ω. What is the probability that the assembly will have a resistance greater than 31 Ω? What is the maximum standard deviation permissible for a probability of 0.01 of the resistance exceeding 31 Ω?

APPENDIX E
USE OF THE WEIBULL DISTRIBUTION

E-1 DATA EVALUATION

The Weibull distribution is used rather widely for evaluating data from various experiments and tests. The equation for the distribution is given by Eq. (A-35). Test data are ranked (Appendix C) and plotted on Weibull paper. The best straight line is drawn through the data points. The lower bound x_0 can be read directly from the plot. The characteristic value θ is the value of the random variable x corresponding to 63.2 percent failure. The slope b is calculated by making linear measurements on the plot without regard to the labels on ordinate and abscissa. In other words, the slope is a geometrical measurement, made as if both axes had linear scales.

The mean of the Weibull distribution is not necessarily the value of the variable corresponding to 50 percent failure. The Weibull distribution is skewed and the mean is a function of the slope as indicated by Eq. (A-38). The relation between the mean (in percent failed) and the slope is shown in Fig. E-1.

It is known that the accuracy of an estimate from test statistics is dependent on the number of items tested. Figure E-2 gives the error in Weibull slope for 50 percent confidence while Fig. E-3 gives the error in slope for 90 percent

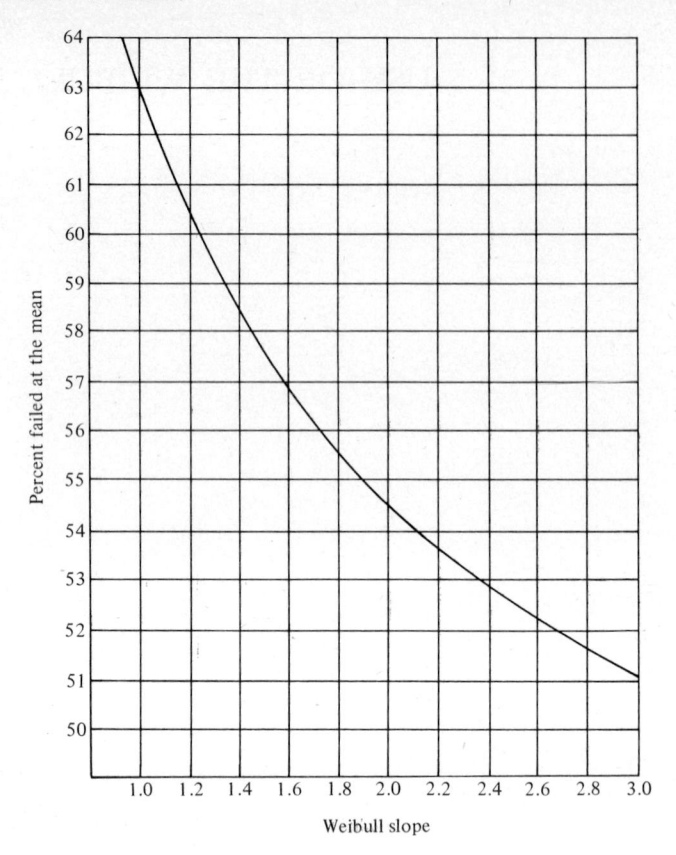

FIGURE E-1
Position of the Weibull mean.

confidence. The plus-minus values on these figures means that the true population slope will be within the plus-minus percent of the observed sample slope with the confidence level indicated.

Figure E-4 gives confidence limits for the true mean as a function of both the sample size and the Weibull slope for 90 percent confidence. Figure E-5 gives similar confidence limits for failure of 10 percent of the population. It is noted that the two sets of curves in Figs. E-4 and E-5 give a number by which the sample value is multiplied. The results indicate the limits within which the true population mean (or 10 percent failure) will be found with 90 percent confidence.

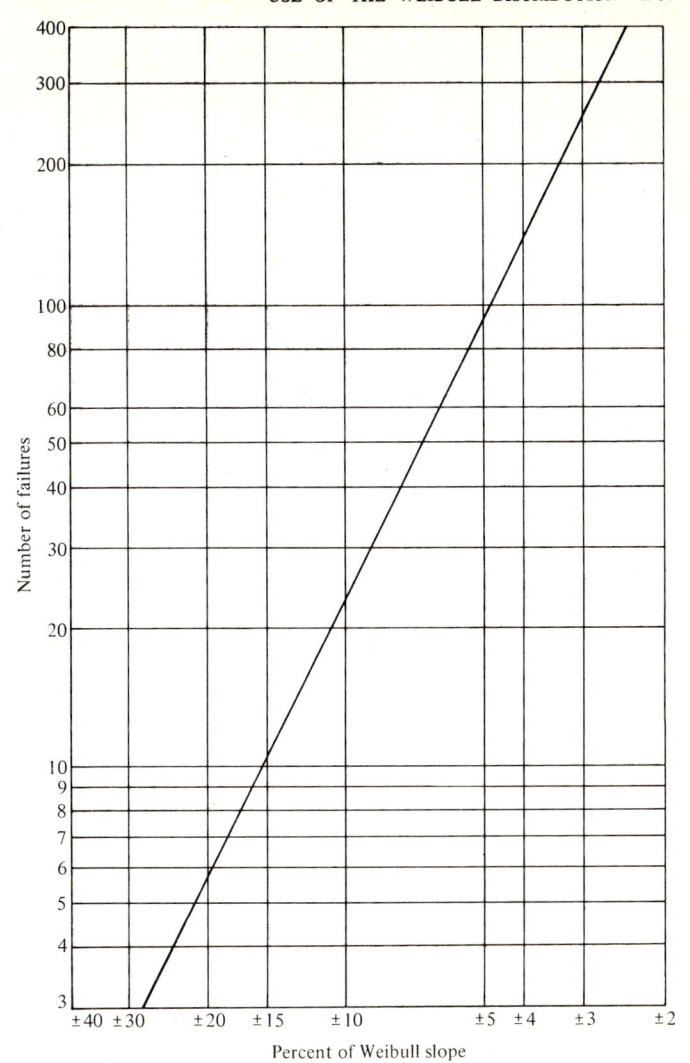

FIGURE E-2
Weibull slope error—50 percent confidence interval.

248 INTRODUCTION TO RELIABILITY IN DESIGN

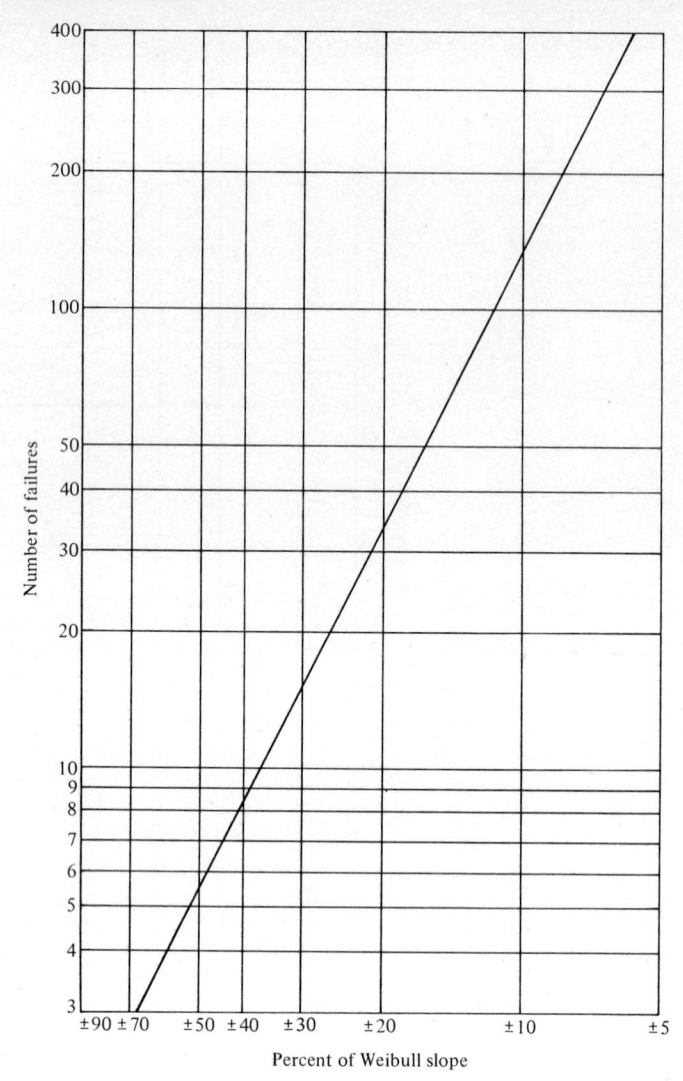

FIGURE E-3
Weibull slope error—90 percent confidence interval.

USE OF THE WEIBULL DISTRIBUTION 249

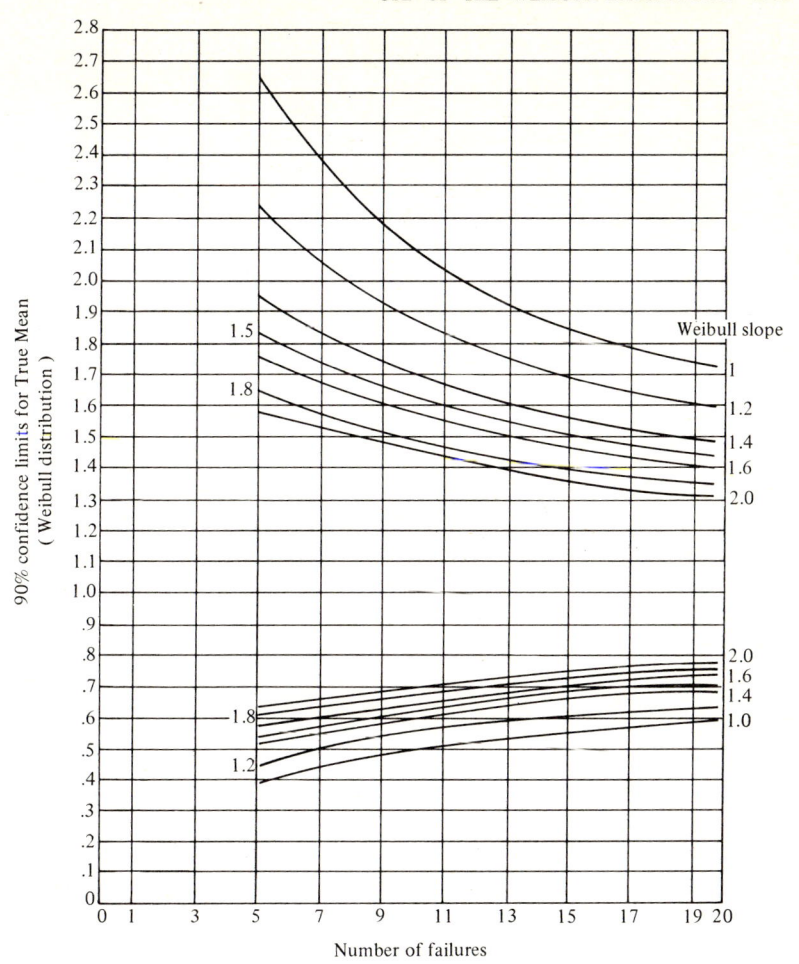

FIGURE E-4
Ninety percent confidence limits for true mean.

250 INTRODUCTION TO RELIABILITY IN DESIGN

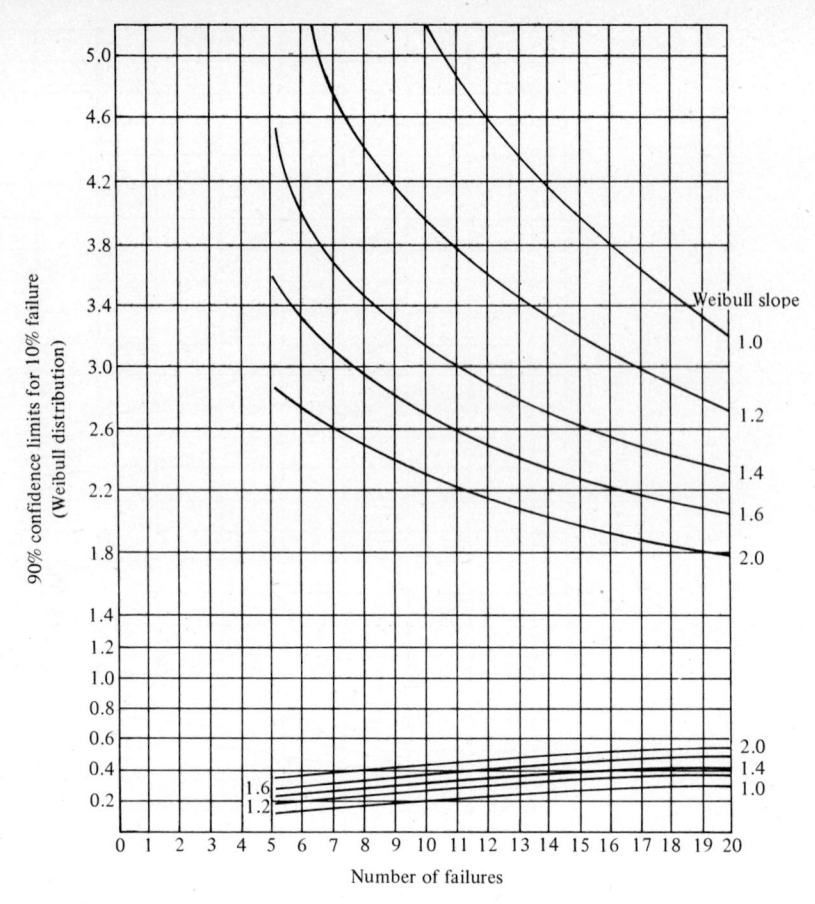

FIGURE E-5
Ninety percent confidence limits for 10 percent failure.

E-2 COMPARISON OF TWO SETS OF TEST STATISTICS

Tests are often made to compare two lots of materials or to compare a given lot with a standard. When the results are plotted, the Weibull slopes are not necessarily equal. In many cases, however, they will be approximately equal. For this situation, the two lots may be compared as follows:

Calculate the degrees of freedom (DF)

Degrees of freedom, lot 1 = $N_1 - 1 = DF_1$
Degrees of freedom, lot 2 = $N_2 - 1 = DF_2$
Total degrees of freedom = $DF_1 \times DF_2 = (N_1 - 1)(N_2 - 1)$

Using the sample Weibull slopes, determine the location of the mean (in percent failed) from Fig. E-1, for each lot. Determine the corresponding value of "life"

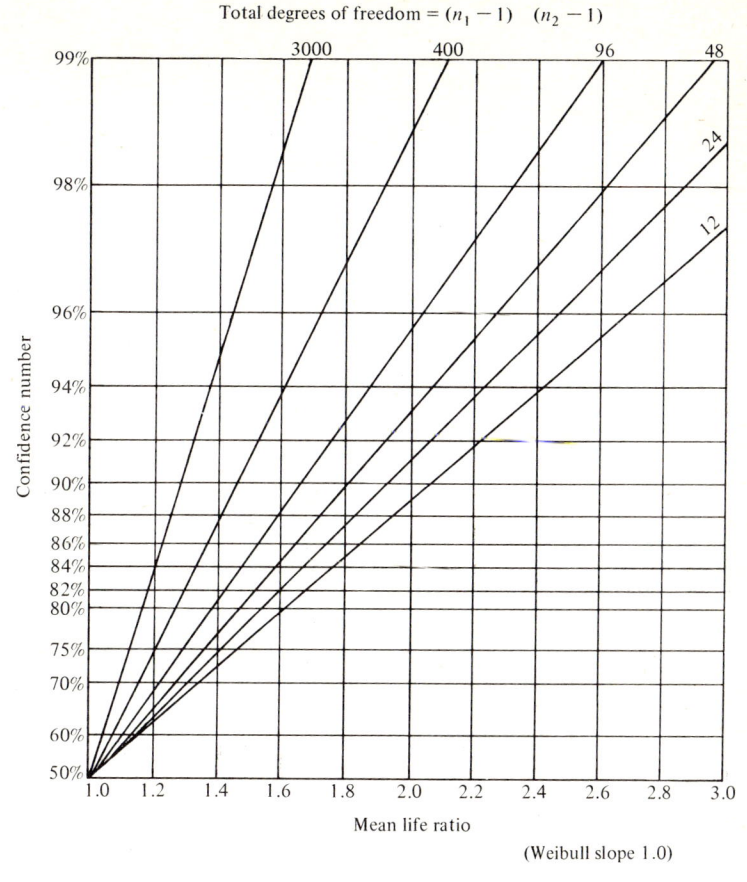

Figure E-6
Confidence numbers for comparing samples; Weibull slope = 1.0.

from the two sample plots. Take the ratio of these two means, dividing the larger by the smaller to get a "mean life ratio." Knowing the mean life ratio, the total degrees of freedom, and the Weibull slope, a confidence number can be obtained from Figs. E-6 through E-11. This confidence number is the probability that the true mean life ratio of the two populations is greater than unity. Interpolation can be used for cases in which the Weibull slope falls between the values for which Figs. E-6 through E-11 were constructed.

For the case of unequal Weibull slopes, the technique is only slightly more involved. The number of degrees of freedom is calculated as above. The slope and mean life of each lot is determined. The mean life ratio is determined as above. Obtain two confidence numbers: one by treating both lots as if both had the slope of lot 1; the second by treating both lots as if both had the slope of lot 2. Average the two confidence numbers and interpret as above for the case of equal Weibull slopes.

252 INTRODUCTION TO RELIABILITY IN DESIGN

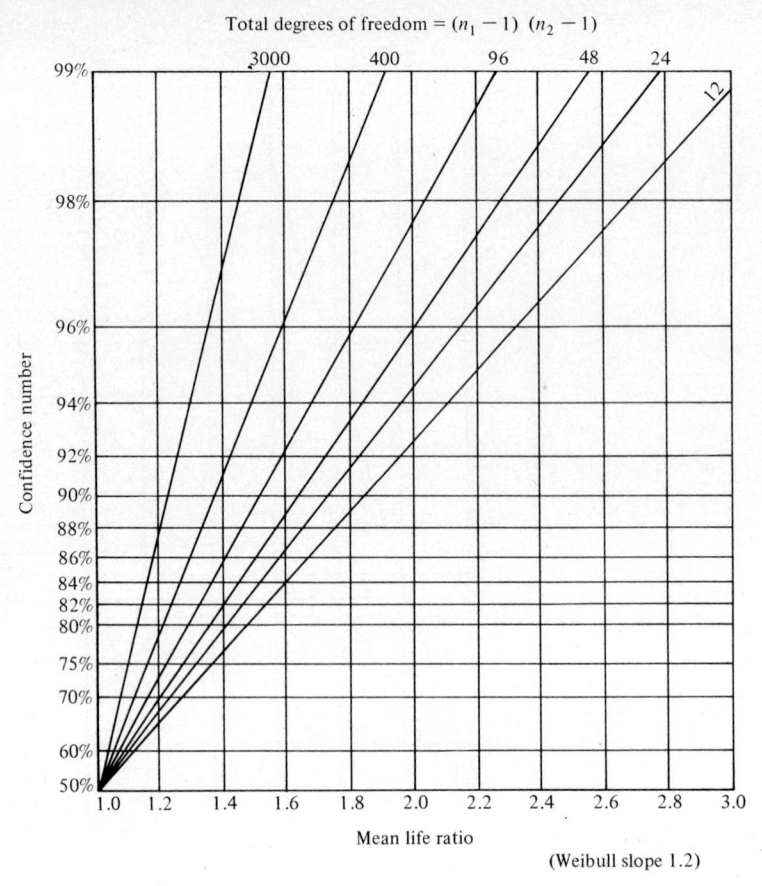

FIGURE E-7
Confidence numbers for comparing samples; Weibull slope = 1.2.

USE OF THE WEIBULL DISTRIBUTION 253

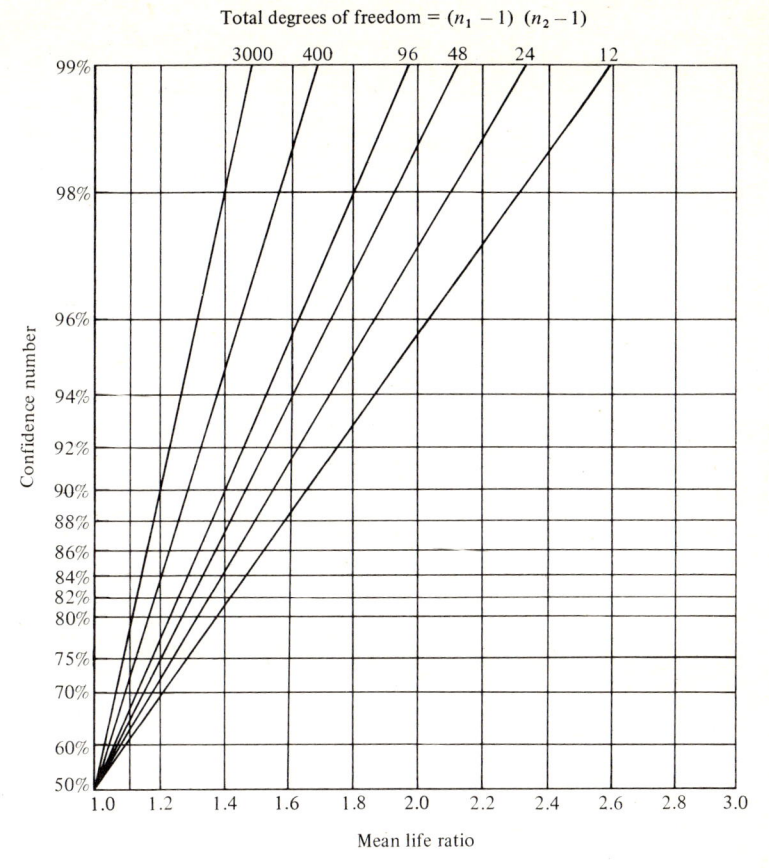

FIGURE E-8
Confidence numbers for comparing samples; Weibull slope = 1.4.

254 INTRODUCTION TO RELIABILITY IN DESIGN

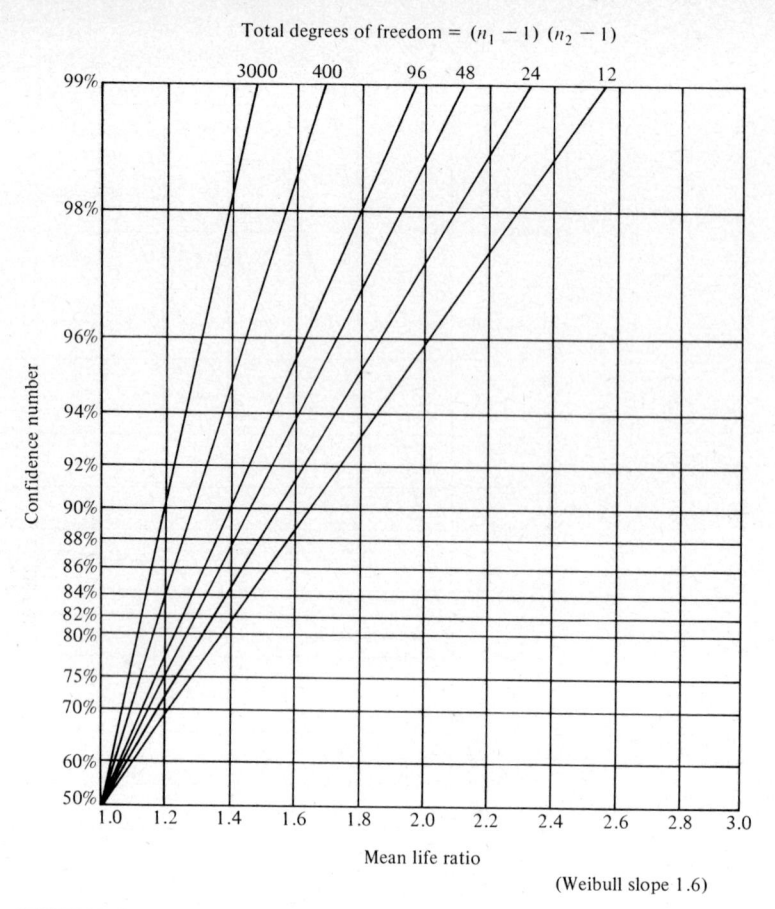

FIGURE E-9
Confidence numbers for comparing samples; Weibull slope = 1.6.

USE OF THE WEIBULL DISTRIBUTION

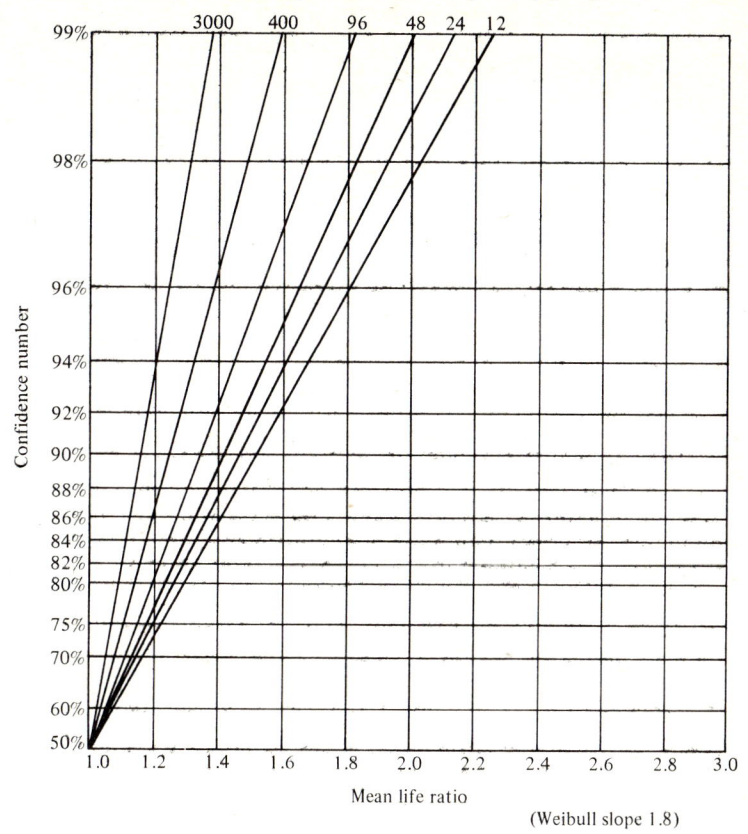

FIGURE E-10
Confidence numbers for comparing samples; Weibull slope = 1.8.

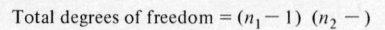

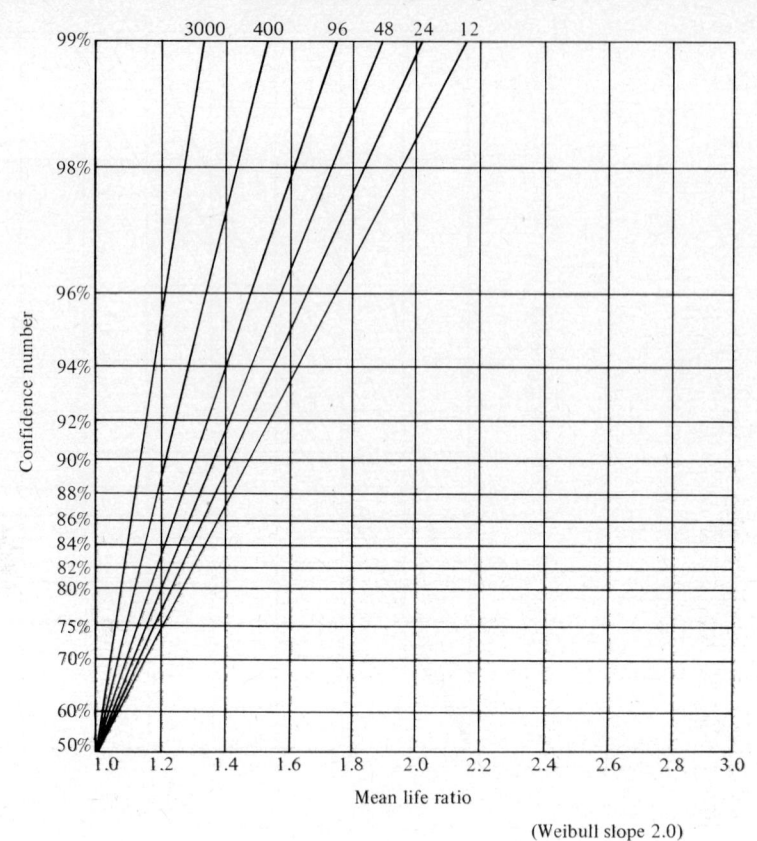

FIGURE E-11
Confidence numbers for comparing samples; Weibull slope = 2.0.

APPENDIX F

Table F-1 THE CUMULATIVE STANDARD NORMAL DISTRIBUTION FUNCTION[a]

$$\Phi(u) = \frac{1}{\sqrt{2\pi}} \int_{-\infty}^{u} e^{-z^2/2} \, dz \text{ for } 0.00 \leq (u = z\sigma) \leq 4.99$$

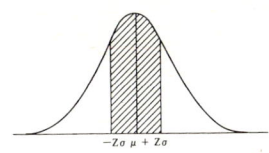

u	0.00	0.01	0.02	0.03	0.04	0.05	0.06	0.07	0.08	0.09
0.0	0.5000	0.5040	0.5080	0.5120	0.5160	0.5199	0.5239	0.5279	0.5319	0.5359
0.1	0.5398	0.5438	0.5478	0.5517	0.5557	0.5596	0.5636	0.5675	0.5714	0.5753
0.2	0.5793	0.5832	0.5871	0.5910	0.5948	0.5987	0.6026	0.6064	0.6103	0.6141
0.3	0.6179	0.6217	0.6255	0.6293	0.6331	0.6368	0.6406	0.6443	0.6480	0.6517
0.4	0.6554	0.6591	0.6628	0.6664	0.6700	0.6736	0.6772	0.6808	0.6844	0.6879
0.5	0.6915	0.6950	0.6985	0.7019	0.7054	0.7088	0.7123	0.7157	0.7190	0.7224
0.6	0.7257	0.7291	0.7324	0.7357	0.7389	0.7422	0.7454	0.7486	0.7517	0.7549
0.7	0.7580	0.7611	0.7642	0.7673	0.7703	0.7734	0.7764	0.7794	0.7823	0.7852
0.8	0.7881	0.7910	0.7939	0.7967	0.7995	0.8023	0.8051	0.8078	0.8106	0.8133
0.9	0.8159	0.8186	0.8212	0.8238	0.8264	0.8289	0.8315	0.8340	0.8365	0.8389
1.0	0.8413	0.8438	0.8461	0.8485	0.8508	0.8531	0.8554	0.8577	0.8599	0.8621
1.1	0.8643	0.8665	0.8686	0.8708	0.8729	0.8749	0.8770	0.8790	0.8810	0.8830
1.2	0.8849	0.8869	0.8888	0.8907	0.8925	0.8944	0.8962	0.8980	0.8997	0.90147
1.3	0.90320	0.90490	0.90658	0.90824	0.90988	0.91149	0.91309	0.91466	0.91621	0.91774
1.4	0.91924	0.92073	0.92220	0.92364	0.92507	0.92647	0.92785	0.92922	0.93056	0.93189
1.5	0.93319	0.93448	0.93574	0.93699	0.93822	0.93943	0.94062	0.94179	0.94295	0.94408
1.6	0.94520	0.94630	0.94738	0.94845	0.94950	0.95053	0.95154	0.95254	0.95352	0.95449
1.7	0.95543	0.95637	0.95728	0.95818	0.95907	0.95994	0.96080	0.96164	0.96246	0.96327
1.8	0.96407	0.96485	0.96562	0.96638	0.96712	0.96784	0.96856	0.96926	0.96995	0.97062
1.9	0.97128	0.97193	0.97257	0.97320	0.97381	0.97441	0.97500	0.97558	0.97615	0.97670
2.0	0.97725	0.97778	0.97831	0.97882	0.97932	0.97982	0.98030	0.98077	0.98124	0.98169
2.1	0.98214	0.98257	0.98300	0.98341	0.98382	0.98422	0.98461	0.98500	0.98537	0.98574
2.2	0.98610	0.98645	0.98679	0.98713	0.98745	0.98778	0.98809	0.98840	0.98870	0.98899
2.3	0.98928	0.98956	0.98983	0.9^20097	0.9^20358	0.9^20613	0.9^20863	0.9^21106	0.9^21344	0.9^21576
2.4	0.9^21802	0.9^22024	0.9^22240	0.9^22451	0.9^22656	0.9^22857	0.9^23053	0.9^23244	0.9^23431	0.9^23613
2.5	0.9^23790	0.9^23963	0.9^24132	0.9^24297	0.9^24457	0.9^24614	0.9^24766	0.9^24915	0.9^25060	0.9^25201
2.6	0.9^25339	0.9^25473	0.9^25604	0.9^25731	0.9^25855	0.9^25975	0.9^26093	0.9^26207	0.9^26319	0.9^26427
2.7	0.9^26533	0.9^26636	0.9^26736	0.9^26833	0.9^26928	0.9^27020	0.9^27110	0.9^27197	0.9^27282	0.9^27365
2.8	0.9^27445	0.9^27523	0.9^27599	0.9^27673	0.9^27744	0.9^27814	0.9^27882	0.9^27948	0.9^28012	0.9^28074
2.9	0.9^28134	0.9^28193	0.9^28250	0.9^28305	0.9^28359	0.9^28411	0.9^28462	0.9^28511	0.9^28559	0.9^28605
3.0	0.9^28650	0.9^28694	0.9^28736	0.9^28777	0.9^28817	0.9^28856	0.9^28893	0.9^28930	0.9^28965	0.9^28999
3.1	0.9^30324	0.9^30646	0.9^30957	0.9^31260	0.9^31553	0.9^31836	0.9^32112	0.9^32378	0.9^32636	0.9^32886
3.2	0.9^33129	0.9^33363	0.9^33590	0.9^33810	0.9^34024	0.9^34230	0.9^34429	0.9^34623	0.9^34810	0.9^34991
3.3	0.9^35166	0.9^35335	0.9^35499	0.9^35658	0.9^35811	0.9^35959	0.9^36103	0.9^36242	0.9^36376	0.9^36505
3.4	0.9^36631	0.9^36752	0.9^36869	0.9^36982	0.9^37091	0.9^37197	0.9^37299	0.9^37398	0.9^37493	0.9^37585
3.5	0.9^37674	0.9^37759	0.9^37842	0.9^37922	0.9^37999	0.9^38074	0.9^38146	0.9^38215	0.9^38282	0.9^38347
3.6	0.9^38409	0.9^38469	0.9^38527	0.9^38583	0.9^38637	0.9^38689	0.9^38739	0.9^38787	0.9^38834	0.9^38879
3.7	0.9^38922	0.9^38964	0.9^40039	0.9^40426	0.9^40799	0.9^41158	0.9^41504	0.9^41838	0.9^42159	0.9^42468
3.8	0.9^42765	0.9^43052	0.9^43327	0.9^43593	0.9^43848	0.9^44094	0.9^44331	0.9^44558	0.9^44777	0.9^44988
3.9	0.9^45190	0.9^45385	0.9^45573	0.9^45753	0.9^45926	0.9^46092	0.9^46253	0.9^46406	0.9^46554	0.9^46696
4.0	0.9^46833	0.9^46964	0.9^47090	0.9^47211	0.9^47327	0.9^47439	0.9^47546	0.9^47649	0.9^47748	0.9^47843
4.1	0.9^47934	0.9^48022	0.9^48106	0.9^48186	0.9^48263	0.9^48338	0.9^48409	0.9^48477	0.9^48542	0.9^48605
4.2	0.9^48665	0.9^48723	0.9^48778	0.9^48832	0.9^48882	0.9^48931	0.9^48978	0.9^50226	0.9^50655	0.9^51066
4.3	0.9^51460	0.9^51837	0.9^52199	0.9^52545	0.9^52876	0.9^53193	0.9^53497	0.9^53788	0.9^54066	0.9^54332
4.4	0.9^54587	0.9^54831	0.9^55065	0.9^55288	0.9^55502	0.9^55706	0.9^55902	0.9^56089	0.9^56268	0.9^56439
4.5	0.9^56602	0.9^56759	0.9^56908	0.9^57051	0.9^57187	0.9^57318	0.9^57442	0.9^57561	0.9^57675	0.9^57784
4.6	0.9^57888	0.9^57987	0.9^58081	0.9^58172	0.9^58258	0.9^58340	0.9^58419	0.9^58494	0.9^58566	0.9^58634
4.7	0.9^58699	0.9^58761	0.9^58821	0.9^58877	0.9^58931	0.9^58983	0.9^60320	0.9^60789	0.9^61235	0.9^61661
4.8	0.9^62067	0.9^62453	0.9^62822	0.9^63173	0.9^63508	0.9^63827	0.9^64131	0.9^64420	0.9^64696	0.9^64958
4.9	0.9^65208	0.9^65446	0.9^65673	0.9^65889	0.9^66094	0.9^66289	0.9^66475	0.9^66652	0.9^66821	0.9^66981

Example: $\Phi(3.57) = 0.9^38215 = 0.9998215$

[a] See Sec. A-10-2d.

APPENDIX G

Table G-1 VALUES OF THE NEGATIVE EXPONENTIAL $R = e^{-F}$ where $F = \lambda t$

F	0.0000[a]	0.0001	0.0002	0.0003	0.0004	0.0005	0.0006	0.0007	0.0008	0.0009
0.000	1.00000	0.99990	0.99980	0.99970	0.99960	0.99950	0.99940	0.99930	0.99920	0.99910
0.001	0.99900	0.99890	0.99880	0.99870	0.99860	0.99850	0.99840	0.99830	0.99820	0.99810
0.002	0.99800	0.99790	0.99780	0.99770	0.99760	0.99750	0.99740	0.99730	0.99720	0.99710
0.003	0.99700	0.99690	0.99681	0.99671	0.99661	0.99651	0.99641	0.99631	0.99621	0.99611
0.004	0.99600	0.99591	0.99581	0.99571	0.99561	0.99551	0.99541	0.99531	0.99521	0.99511
0.005	0.99501	0.99491	0.99481	0.99471	0.99461	0.99452	0.99442	0.99432	0.99422	0.99412
0.006	0.99402	0.99392	0.99382	0.99372	0.99362	0.99352	0.99342	0.99332	0.99322	0.99312
0.007	0.99302	0.99293	0.99283	0.99273	0.99263	0.99253	0.99243	0.99233	0.99223	0.99213
0.008	0.99203	0.99193	0.99183	0.99173	0.99164	0.99154	0.99144	0.99134	0.99124	0.99114
0.009	0.99104	0.99094	0.99084	0.99074	0.99064	0.99054	0.99045	0.99035	0.99025	0.99015
0.010	0.99005	0.98995	0.98985	0.98975	0.98965	0.98955	0.98946	0.98936	0.98926	0.98916
0.011	0.98906	0.98896	0.98886	0.98876	0.98866	0.98857	0.98847	0.98837	0.98827	0.98817
0.012	0.98807	0.98797	0.98787	0.98778	0.98768	0.98758	0.98748	0.98738	0.98728	0.98718
0.013	0.98708	0.98699	0.98689	0.98679	0.98669	0.98659	0.98649	0.98639	0.98629	0.98620
0.014	0.98610	0.98600	0.98590	0.98580	0.98570	0.98560	0.98551	0.98541	0.98531	0.98521
0.015	0.98511	0.98501	0.98491	0.98482	0.98472	0.98462	0.98452	0.98442	0.98432	0.98423
0.016	0.98413	0.98403	0.98393	0.98383	0.98373	0.98364	0.98354	0.98344	0.98334	0.98324
0.017	0.98314	0.98305	0.98295	0.98285	0.98275	0.98265	0.98255	0.98246	0.98236	0.98226
0.018	0.98216	0.98206	0.98196	0.98187	0.98177	0.98167	0.98157	0.98147	0.98138	0.98128
0.019	0.98118	0.98108	0.98098	0.98089	0.98079	0.98069	0.98059	0.98049	0.98039	0.98030
0.020	0.98020	0.98010	0.98000	0.97990	0.97981	0.97971	0.97961	0.97951	0.97941	0.97932
0.021	0.97922	0.97912	0.97902	0.97893	0.97883	0.97873	0.97863	0.97853	0.97844	0.97834
0.022	0.97824	0.97814	0.97804	0.97795	0.97785	0.97775	0.97765	0.97756	0.97746	0.97736
0.023	0.97726	0.97716	0.97707	0.97697	0.97687	0.97677	0.97668	0.97658	0.97648	0.97638
0.024	0.97629	0.97619	0.97609	0.97599	0.97590	0.97580	0.97570	0.97560	0.97550	0.97541
0.025	0.97531	0.97521	0.97511	0.97502	0.97492	0.97482	0.97472	0.97463	0.97453	0.97443
0.026	0.97434	0.97424	0.97414	0.97404	0.97395	0.97385	0.97375	0.97365	0.97356	0.97346
0.027	0.97336	0.97326	0.97317	0.97307	0.97297	0.97287	0.97278	0.97268	0.97258	0.97249
0.028	0.97239	0.97229	0.97219	0.97210	0.97200	0.97190	0.97181	0.97171	0.97161	0.97151
0.029	0.97142	0.97132	0.97122	0.97113	0.97103	0.97093	0.97083	0.97074	0.97064	0.97054
0.030	0.97045	0.97035	0.97025	0.97015	0.97006	0.96996	0.96986	0.96977	0.96967	0.96957
0.031	0.96948	0.96938	0.96928	0.96918	0.96909	0.96899	0.96889	0.96880	0.96870	0.96860
0.032	0.96851	0.96841	0.96831	0.96822	0.96812	0.96802	0.96793	0.96783	0.96773	0.96764
0.033	0.96754	0.96744	0.96735	0.96725	0.96715	0.96705	0.96696	0.96686	0.96676	0.96667
0.034	0.96657	0.96647	0.96638	0.96628	0.96618	0.96609	0.96599	0.96590	0.96580	0.96570
0.035	0.96561	0.96551	0.96541	0.96532	0.96522	0.96512	0.96503	0.96493	0.96483	0.96474
0.036	9.96464	0.96454	0.96445	0.96435	0.96425	0.96416	0.96406	0.96397	0.96387	0.96377
0.037	0.96368	0.96358	0.96348	0.96339	0.96329	0.96319	0.96310	0.96300	0.96291	0.96281
0.038	0.96271	0.96262	0.96252	0.96242	0.96233	0.96223	0.96214	0.96204	0.96194	0.96185
0.039	0.96175	0.96165	0.96156	0.96146	0.96137	0.96127	0.96117	0.96108	0.96098	0.96089
0.040	0.96079	0.96069	0.96060	0.96050	0.96041	0.96031	0.96021	0.96012	0.96002	0.95993
0.041	0.95983	0.95973	0.95964	0.95954	0.95945	0.95935	0.95925	0.95916	0.95906	0.95897
0.042	0.95887	0.95877	0.95868	0.95858	0.95849	0.95839	0.95829	0.95820	0.95810	0.95801
0.043	0.95791	0.95782	0.95772	0.95762	0.95753	0.95743	0.95734	0.95724	0.95715	0.95705
0.044	0.95695	0.95686	0.95676	0.95667	0.95657	0.95648	0.95638	0.95628	0.95619	0.95609

[a] Column headings provide fourth decimal place for value of F.

Table G-1 (continued)

F	0.0000[a]	0.0001	0.0002	0.0003	0.0004	0.0005	0.0006	0.0007	0.0008	0.0009
0.045	0.95600	0.95590	0.95581	0.95571	0.95562	0.95552	0.95542	0.95533	0.95523	0.95514
0.046	0.95504	0.95495	0.95485	0.95476	0.95466	0.95456	0.95447	0.95437	0.95428	0.95418
0.047	0.95409	0.95399	0.95390	0.95380	0.95371	0.95361	0.95352	0.95342	0.95332	0.95323
0.048	0.95313	0.95304	0.95294	0.95285	0.95275	0.95266	0.95256	0.95247	0.95237	0.95228
0.049	0.95218	0.95209	0.95199	0.95190	0.95180	0.95171	0.95161	0.95151	0.95142	0.95132
0.050	0.95123	0.95113	0.95104	0.95094	0.95085	0.95075	0.95066	0.95056	0.95047	0.95037
0.051	0.95028	0.95018	0.95009	0.94999	0.94990	0.94980	0.94971	0.94961	0.94952	0.94942
0.052	0.94933	0.94923	0.94914	0.94904	0.94895	0.94885	0.94876	0.94866	0.94857	0.94847
0.053	0.94838	0.94829	0.94819	0.94810	0.94800	0.94791	0.94781	0.94772	0.94762	0.94753
0.054	0.94743	0.94734	0.94724	0.94715	0.94705	0.94696	0.94686	0.94677	0.94667	0.94658
0.055	0.94649	0.94639	0.94630	0.94620	0.94611	0.94601	0.94592	0.94582	0.94573	0.94563
0.056	0.94554	0.94544	0.94535	0.94526	0.94516	0.94507	0.94497	0.94488	0.94478	0.94469
0.057	0.94459	0.94450	0.94441	0.94431	0.94422	0.94412	0.94403	0.94393	0.94384	0.94374
0.058	0.94365	0.94356	0.94346	0.94337	0.94327	0.94318	0.94308	0.94299	0.94290	0.94280
0.059	0.94271	0.94261	0.94252	0.94242	0.94233	0.94224	0.94214	0.94205	0.94195	0.94186
0.060	0.94176	0.94167	0.94158	0.94148	0.94139	0.94129	0.94120	0.94111	0.94101	0.94092
0.061	0.94082	0.94073	0.94064	0.94054	0.94045	0.94035	0.94026	0.94016	0.94007	0.93998
0.062	0.93988	0.93979	0.93969	0.93960	0.93951	0.93941	0.93932	0.93923	0.93913	0.93904
0.063	0.93894	0.93885	0.93876	0.93866	0.93857	0.93847	0.93838	0.93829	0.93819	0.93810
0.064	0.93800	0.93791	0.93782	0.93772	0.93763	0.93754	0.93744	0.93735	0.93725	0.93716
0.065	0.93707	0.93697	0.93688	0.93679	0.93669	0.93660	0.93651	0.93641	0.93632	0.93622
0.066	0.93613	0.93604	0.93594	0.93585	0.93576	0.93566	0.93557	0.93548	0.93538	0.93529
0.067	0.93520	0.93510	0.93501	0.93491	0.93482	0.93473	0.93463	0.93454	0.93445	0.93435
0.068	0.93426	0.93417	0.93407	0.93398	0.93389	0.93379	0.93370	0.93361	0.93351	0.93342
0.069	0.93333	0.93323	0.93314	0.93305	0.93295	0.93286	0.93277	0.93267	0.93258	0.93249
0.070	0.93239	0.93230	0.93221	0.93211	0.93202	0.93193	0.93183	0.93174	0.93165	0.93156
0.071	0.93146	0.93137	0.93128	0.93118	0.93109	0.93100	0.93090	0.93081	0.93072	0.93062
0.072	0.93053	0.93044	0.93034	0.93025	0.93016	0.93007	0.92997	0.92988	0.92979	0.92969
0.073	0.92960	0.92951	0.92941	0.92932	0.92923	0.92914	0.92904	0.92895	0.92886	0.92876
0.074	0.92867	0.92858	0.92849	0.92839	0.92830	0.92821	0.92811	0.92802	0.92793	0.92784
0.075	0.92774	0.92765	0.92756	0.92747	0.92737	0.92728	0.92719	0.92709	0.92700	0.92691
0.076	0.92682	0.92672	0.92663	0.92654	0.92645	0.92635	0.92626	0.92617	0.92608	0.92598
0.077	0.92589	0.92580	0.92570	0.92561	0.92552	0.92543	0.92533	0.92524	0.92515	0.92506
0.078	0.92496	0.92487	0.92478	0.92469	0.92459	0.92450	0.92441	0.92432	0.92422	0.92413
0.079	0.92404	0.92395	0.92386	0.92376	0.92367	0.92358	0.92349	0.92339	0.92330	0.92321
0.080	0.92312	0.92302	0.92293	0.92284	0.92275	0.92265	0.92256	0.92247	0.92238	0.92229
0.081	0.92219	0.92210	0.92201	0.92192	0.92182	0.92173	0.92164	0.92155	0.92146	0.92136
0.082	0.92127	0.92118	0.92109	0.92100	0.92090	0.92081	0.92072	0.92063	0.92054	0.92044
0.083	0.92035	0.92026	0.92017	0.92008	0.91998	0.91989	0.91980	0.91971	0.91962	0.91952
0.084	0.91943	0.91934	0.91925	0.91916	0.91906	0.91897	0.91888	0.91879	0.91870	0.91860
0.085	0.91851	0.91842	0.91833	0.91824	0.91814	0.91805	0.91796	0.91787	0.91778	0.91769
0.086	0.91759	0.91750	0.91741	0.91732	0.91723	0.91714	0.91704	0.91695	0.91686	0.91677
0.087	0.91668	0.91659	0.91649	0.91640	0.91631	0.91622	0.91613	0.91604	0.91594	0.91585
0.088	0.91576	0.91567	0.91558	0.91549	0.91539	0.91530	0.91521	0.91512	0.91503	0.91494
0.089	0.91485	0.91475	0.91466	0.91457	0.91448	0.91439	0.91430	0.91421	0.91411	0.91402

[a] Column headings provide fourth decimal place for value of F.

Table G-1 (continued)

F	0.0000[a]	0.0001	0.0002	0.0003	0.0004	0.0005	0.0006	0.0007	0.0008	0.0009
0.090	0.91393	0.91384	0.91375	0.91366	0.91357	0.91347	0.91338	0.91329	0.91320	0.91311
0.091	0.91302	0.91293	0.91284	0.91274	0.91265	0.91256	0.91247	0.91238	0.91229	0.91220
0.092	0.91211	0.91201	0.91192	0.91183	0.91174	0.91165	0.91156	0.91147	0.91138	0.91128
0.093	0.91119	0.91110	0.91101	0.91092	0.91083	0.91074	0.91065	0.91056	0.91046	0.91037
0.094	0.91028	0.91019	0.91010	0.91001	0.90992	0.90983	0.90974	0.90965	0.90955	0.90946
0.095	0.90937	0.90928	0.90919	0.90910	0.90901	0.90892	0.90883	0.90874	0.90865	0.90855
0.096	0.90846	0.90837	0.90828	0.90819	0.90810	0.90801	0.90792	0.90783	0.90774	0.90765
0.097	0.90756	0.90747	0.90737	0.90728	0.90719	0.90710	0.90701	0.90692	0.90683	0.90674
0.098	0.90665	0.90656	0.90647	0.90638	0.90629	0.90620	0.90611	0.90601	0.90592	0.90583
0.099	0.90574	0.90565	0.90556	0.90547	0.90538	0.90529	0.90520	0.90511	0.90502	0.90493

F	0.000[b]	0.001	0.002	0.003	0.004	0.005	0.006	0.007	0.008	0.009
0.10	0.90484	0.90393	0.90303	0.90213	0.90123	0.90032	0.89942	0.89853	0.89763	0.89673
0.11	0.89583	0.89494	0.89404	0.89315	0.89226	0.89137	0.89048	0.88959	0.88870	0.88781
0.12	0.88692	0.88603	0.88515	0.88426	0.88338	0.88250	0.88161	0.88073	0.87985	0.87897
0.13	0.87810	0.87722	0.87634	0.87547	0.87459	0.87372	0.87284	0.87197	0.87110	0.87023
0.14	0.86936	0.86849	0.86762	0.86675	0.86589	0.86502	0.86416	0.86329	0.86243	0.86157
0.15	0.86071	0.85985	0.85899	0.85813	0.85727	0.85642	0.85556	0.85470	0.85385	0.85300
0.16	0.85214	0.85129	0.85044	0.84959	0.84874	0.84789	0.84705	0.84620	0.84535	0.84451
0.17	0.84366	0.84282	0.84198	0.84114	0.84030	0.83946	0.83862	0.83778	0.83694	0.83611
0.18	0.83527	0.83444	0.83360	0.83277	0.83194	0.83110	0.83027	0.82944	0.82861	0.82779
0.19	0.82696	0.82613	0.82531	0.82448	0.82366	0.82283	0.82201	0.82119	0.82037	0.81955
0.20	0.81873	0.81791	0.81709	0.81628	0.81546	0.81465	0.81383	0.81302	0.81221	0.81140
0.21	0.81058	0.80977	0.80896	0.80816	0.80735	0.80654	0.80574	0.80493	0.80413	0.80332
0.22	0.80252	0.80172	0.80092	0.80011	0.79932	0.79852	0.79772	0.79692	0.79612	0.79533
0.23	0.79453	0.79374	0.79295	0.79215	0.79136	0.79057	0.78978	0.78899	0.78820	0.78741
0.24	0.78663	0.78584	0.78506	0.78427	0.78349	0.78270	0.78192	0.78114	0.78036	0.77958
0.25	0.77880	0.77802	0.77724	0.77647	0.77569	0.77492	0.77414	0.77337	0.77260	0.77182
0.26	0.77105	0.77028	0.76951	0.76874	0.76797	0.76721	0.76644	0.76567	0.76491	0.76414
0.27	0.76338	0.76262	0.76185	0.76109	0.76033	0.75957	0.75881	0.75805	0.75730	0.75654
0.28	0.75578	0.75503	0.75427	0.75352	0.75277	0.75201	0.75126	0.75051	0.74976	0.74901
0.29	0.74826	0.74752	0.74677	0.74602	0.74528	0.74453	0.74379	0.74304	0.74230	0.74156
0.30	0.74082	0.74008	0.73934	0.73860	0.73786	0.73712	0.73639	0.73565	0.73492	0.73418
0.31	0.73345	0.73271	0.73198	0.73125	0.73052	0.72979	0.72906	0.72833	0.72760	0.72688
0.32	0.72615	0.72542	0.72470	0.72397	0.72325	0.72253	0.72181	0.72108	0.72036	0.71964
0.33	0.71892	0.71821	0.71749	0.71677	0.71605	0.71534	0.71462	0.71391	0.71320	0.71248
0.34	0.71177	0.71106	0.71035	0.70964	0.70893	0.70822	0.70751	0.70681	0.70610	0.70539
0.35	0.70469	0.70398	0.70328	0.70258	0.70187	0.70117	0.70047	0.69977	0.69907	0.69837
0.36	0.69768	0.69698	0.69628	0.69559	0.69489	0.69420	0.69350	0.69281	0.69212	0.69143
0.37	0.69073	0.69004	0.68935	0.68867	0.68798	0.68729	0.68660	0.68592	0.68523	0.68455
0.38	0.68386	0.68318	0.68250	0.68181	0.68113	0.68045	0.67977	0.67909	0.67841	0.67773
0.39	0.67706	0.67638	0.67570	0.67503	0.67435	0.67368	0.67301	0.67233	0.67166	0.67099

[a] Column headings provide fourth decimal place for value of F.
[b] Column headings provide third decimal place for value of F.

Table G-1 (continued)

F	0.000[a]	0.001	0.002	0.003	0.004	0.005	0.006	0.007	0.008	0.009
0.40	0.67032	0.66965	0.66898	0.66831	0.66764	0.66698	0.66631	0.66564	0.66498	0.66431
0.41	0.66365	0.66299	0.66232	0.66166	0.66100	0.66034	0.65968	0.65902	0.65836	0.65770
0.42	0.65705	0.65639	0.65573	0.65508	0.65442	0.65377	0.65312	0.65246	0.65181	0.65116
0.43	0.65051	0.64986	0.64921	0.64856	0.64791	0.64726	0.64662	0.64597	0.64533	0.64468
0.44	0.64404	0.64339	0.64275	0.64211	0.64147	0.64082	0.64018	0.63954	0.63890	0.63827
0.45	0.63763	0.63699	0.63635	0.63572	0.63508	0.63445	0.63381	0.63318	0.63255	0.63192
0.46	0.63128	0.63065	0.63002	0.62939	0.62876	0.62814	0.62751	0.62688	0.62625	0.62563
0.47	0.62500	0.62438	0.62375	0.62313	0.62251	0.62189	0.62126	0.62064	0.62002	0.61940
0.48	0.61878	0.61816	0.61755	0.61693	0.61631	0.61570	0.61508	0.61447	0.61385	0.61324
0.49	0.61263	0.61201	0.61140	0.61079	0.61018	0.60957	0.60896	0.60835	0.60774	0.60714
0.50	0.60653	0.60592	0.60531	0.60471	0.60411	0.60351	0.60290	0.60230	0.60170	0.60110
0.51	0.60050	0.59990	0.59930	0.59870	0.59810	0.59750	0.59690	0.59631	0.59571	0.59512
0.52	0.59452	0.59393	0.59333	0.59274	0.59215	0.59156	0.59096	0.59037	0.58978	0.58919
0.53	0.58860	0.58802	0.58743	0.58684	0.58626	0.58567	0.58508	0.58450	0.58391	0.58333
0.54	0.58275	0.58217	0.58158	0.58100	0.58042	0.57984	0.57926	0.57868	0.57810	0.57753
0.55	0.57695	0.57637	0.57580	0.57522	0.57465	0.57407	0.57350	0.57293	0.57235	0.57178
0.56	0.57121	0.57064	0.57007	0.56950	0.56893	0.56836	0.56779	0.56722	0.56666	0.56609
0.57	0.56553	0.56496	0.56440	0.56383	0.56327	0.56270	0.56214	0.56158	0.56102	0.56046
0.58	0.55990	0.55934	0.55878	0.55822	0.55766	0.55711	0.55655	0.55599	0.55544	0.55488
0.59	0.55433	0.55377	0.55322	0.55267	0.55211	0.55156	0.55101	0.55046	0.54991	0.54936
0.60	0.54881	0.54826	0.54772	0.54718	0.54662	0.54607	0.54553	0.54498	0.54444	0.54389
0.61	0.54335	0.54281	0.54227	0.54172	0.54118	0.54064	0.54010	0.53956	0.53902	0.53848
0.62	0.53794	0.53741	0.53687	0.53633	0.53580	0.53526	0.53473	0.53419	0.53366	0.53312
0.63	0.53259	0.53206	0.53153	0.53100	0.53047	0.52994	0.52941	0.52888	0.52835	0.52782
0.64	0.52729	0.52677	0.52624	0.52571	0.52519	0.52466	0.52414	0.52361	0.52309	0.52257
0.65	0.52205	0.52152	0.52100	0.52048	0.51996	0.51944	0.51892	0.51840	0.51789	0.51737
0.66	0.51685	0.51633	0.51582	0.51530	0.51479	0.51427	0.51376	0.51325	0.51273	0.51222
0.67	0.51171	0.51120	0.51069	0.51018	0.50967	0.50916	0.50865	0.50814	0.50763	0.50712
0.68	0.50662	0.50611	0.50560	0.50510	0.50459	0.50409	0.50359	0.50308	0.50258	0.50208
0.69	0.50158	0.50107	0.50057	0.50007	0.49957	0.49907	0.49858	0.49808	0.49758	0.49708
0.70	0.49659	0.49609	0.49559	0.49510	0.49460	0.49411	0.49361	0.49312	0.49263	0.49214
0.71	0.49164	0.49115	0.49066	0.49017	0.48968	0.48919	0.48870	0.48821	0.48773	0.48724
0.72	0.48675	0.48627	0.48578	0.48529	0.48481	0.48432	0.48384	0.48336	0.48287	0.48239
0.73	0.48191	0.48143	0.48095	0.48047	0.47999	0.47951	0.47903	0.47855	0.47807	0.47759
0.74	0.47711	0.47664	0.47616	0.47568	0.47521	0.47473	0.47426	0.47379	0.47331	0.47284
0.75	0.47237	0.47189	0.47142	0.47095	0.47048	0.47001	0.46954	0.46907	0.46860	0.46813
0.76	0.46767	0.46720	0.46673	0.46627	0.46580	0.46533	0.46487	0.46440	0.46394	0.46348
0.77	0.46301	0.46255	0.46209	0.46163	0.46116	0.46070	0.46024	0.45978	0.45932	0.45886
0.78	0.45841	0.45795	0.45749	0.45703	0.45658	0.45612	0.45566	0.45521	0.45475	0.45430
0.79	0.45384	0.45339	0.45294	0.45249	0.45203	0.45158	0.45113	0.45068	0.45023	0.44978
0.80	0.44933	0.44888	0.44843	0.44798	0.44754	0.44709	0.44664	0.44619	0.44575	0.44530
0.81	0.44486	0.44441	0.44397	0.44353	0.44308	0.44264	0.44220	0.44175	0.44131	0.44087
0.82	0.44043	0.43999	0.43955	0.43911	0.43867	0.43823	0.43780	0.43736	0.43692	0.43649
0.83	0.43605	0.43561	0.43518	0.43474	0.43431	0.43387	0.43344	0.43301	0.43257	0.43214
0.84	0.43171	0.43128	0.43085	0.43042	0.42999	0.42956	0.42913	0.42870	0.42827	0.42784

[a] Column headings provide third decimal place for value of F.

Table G-1 (continued)

F	0.000[a]	0.001	0.002	0.003	0.004	0.005	0.006	0.007	0.008	0.009
0.85	0.42741	0.42699	0.42656	0.42613	0.42571	0.42528	0.42486	0.42443	0.42401	0.42359
0.86	0.42316	0.42274	0.42232	0.42189	0.42147	0.42105	0.42063	0.42021	0.41979	0.41937
0.87	0.41895	0.41853	0.41811	0.41770	0.41728	0.41686	0.41645	0.41603	0.41561	0.41520
0.88	0.41478	0.41437	0.41395	0.41354	0.41313	0.41271	0.41230	0.41189	0.41148	0.41107
0.89	0.41066	0.41025	0.40984	0.40943	0.40902	0.40861	0.40820	0.40779	0.40738	0.40698
0.90	0.40657	0.40616	0.40576	0.40535	0.40495	0.40454	0.40414	0.40373	0.40333	0.40293
0.91	0.40252	0.40212	0.40172	0.40132	0.40092	0.40052	0.40012	0.39972	0.39932	0.39892
0.92	0.39851	0.39812	0.39772	0.39733	0.39693	0.39653	0.39614	0.39574	0.39534	0.39495
0.93	0.39455	0.39416	0.39377	0.39337	0.39298	0.39259	0.39219	0.39180	0.39141	0.39102
0.94	0.39063	0.39024	0.38985	0.38946	0.38907	0.38868	0.38829	0.38790	0.38752	0.38713
0.95	0.38674	0.38635	0.38597	0.38558	0.38520	0.38481	0.38443	0.38404	0.38366	0.38328
0.96	0.38289	0.38251	0.38213	0.38175	0.38136	0.38098	0.38060	0.38022	0.37984	0.37946
0.97	0.37908	0.37870	0.37833	0.37795	0.37757	0.37719	0.37682	0.37644	0.37606	0.37569
0.98	0.37531	0.37494	0.37456	0.37419	0.37381	0.37344	0.37307	0.37269	0.37232	0.37195
0.99	0.37158	0.37121	0.37083	0.37046	0.37009	0.36972	0.36935	0.36898	0.36862	0.36825
1.00	0.36788	0.36751	0.36714	0.36678	0.36641	0.36604	0.36568	0.36531	0.36495	0.36458
1.01	0.36422	0.36385	0.36349	0.36313	0.36277	0.36240	0.36204	0.36168	0.36132	0.36096
1.02	0.36059	0.36023	0.35987	0.35951	0.35916	0.35880	0.35844	0.35808	0.35772	0.35736
1.03	0.35701	0.35665	0.35629	0.35594	0.35558	0.35523	0.35487	0.35452	0.35416	0.35381
1.04	0.35345	0.35310	0.35275	0.35240	0.35204	0.35169	0.35134	0.35099	0.35064	0.35029
1.05	0.34994	0.34959	0.34924	0.34889	0.34854	0.34819	0.34784	0.34750	0.34715	0.34680
1.06	0.34646	0.34611	0.34576	0.34542	0.34507	0.34473	0.34438	0.34404	0.34370	0.34335
1.07	0.34301	0.34267	0.34232	0.34198	0.34164	0.34130	0.34096	0.34062	0.34028	0.33994
1.08	0.33960	0.33926	0.33892	0.33858	0.33824	0.33790	0.33756	0.33723	0.33689	0.33655
1.09	0.33622	0.33588	0.33554	0.33521	0.33487	0.33454	0.33421	0.33387	0.33354	0.33320

F	0.00[b]	0.01	0.02	0.03	0.04	0.05	0.06	0.07	0.08	0.09
1.1	0.33287	0.32956	0.32628	0.32303	0.31982	0.31664	0.31349	0.31037	0.30728	0.30422
1.2	0.30119	0.29820	0.29523	0.29229	0.28938	0.28650	0.28365	0.28083	0.27804	0.27527
1.3	0.27253	0.26982	0.26714	0.26448	0.26185	0.25924	0.25666	0.25411	0.25158	0.24908
1.4	0.24660	0.24414	0.24171	0.23931	0.23693	0.23457	0.23224	0.22992	0.22764	0.22537
1.5	0.22313	0.22090	0.21871	0.21654	0.21438	0.21225	0.21014	0.20805	0.20598	0.20393
1.6	0.20190	0.19989	0.19790	0.19593	0.19398	0.19205	0.19014	0.18825	0.18637	0.18452
1.7	0.18268	0.18087	0.17907	0.17728	0.17552	0.17377	0.17204	0.17033	0.16864	0.16696
1.8	0.16530	0.16365	0.16203	0.16041	0.15882	0.15724	0.15567	0.15412	0.15259	0.15107
1.9	0.14957	0.14808	0.14661	0.14515	0.14370	0.14227	0.14086	0.13946	0.13807	0.13669
2.0	0.13533	0.13399	0.13266	0.13134	0.13003	0.12873	0.12745	0.12619	0.12493	0.12369
2.1	0.12246	0.12124	0.12003	0.11884	0.11765	0.11648	0.11533	0.11418	0.11304	0.11192
2.2	0.11080	0.10970	0.10861	0.10753	0.10645	0.10540	0.10435	0.10331	0.10228	0.10127
2.3	0.10030	0.09926	0.09827	0.09730	0.09633	0.09537	0.09442	0.09348	0.09255	0.09163
2.4	0.09072	0.08982	0.08892	0.08804	0.08716	0.08629	0.08543	0.08458	0.08374	0.08291
2.5	0.08209	0.08127	0.08046	0.07966	0.07887	0.07808	0.07731	0.07654	0.07577	0.07502

[a] Column headings provide third decimal place for value of F.
[b] Column headings provide second decimal place for value of F.

Table G-1 (continued)

F	0.00[a]	0.01	0.02	0.03	0.04	0.05	0.06	0.07	0.08	0.09
2.6	0.07427	0.07353	0.07280	0.07208	0.07136	0.07065	0.06995	0.06925	0.06856	0.06788
2.7	0.06721	0.06654	0.06587	0.06522	0.06457	0.06393	0.06329	0.06266	0.06204	0.06142
2.8	0.06081	0.06020	0.05960	0.05901	0.05843	0.05784	0.05727	0.05670	0.05614	0.05558
2.9	0.05502	0.05448	0.05393	0.05340	0.05287	0.05234	0.05182	0.05130	0.05079	0.05029
3.0	0.04979	0.04929	0.04880	0.04832	0.04784	0.04736	0.04689	0.04642	0.04596	0.04550
3.1	0.04505	0.04460	0.04416	0.04372	0.04328	0.04285	0.04243	0.04200	0.04159	0.04117
3.2	0.04076	0.04036	0.03996	0.03956	0.03916	0.03877	0.03839	0.03800	0.03763	0.03725
3.3	0.03688	0.03652	0.03615	0.03579	0.03544	0.03508	0.03473	0.03439	0.03405	0.03371
3.4	0.03337	0.03304	0.03271	0.03239	0.03206	0.03175	0.03143	0.03112	0.03081	0.03050
3.5	0.03020	0.02990	0.02960	0.02930	0.02901	0.02872	0.02844	0.02816	0.02788	0.02760
3.6	0.02732	0.02705	0.02678	0.02652	0.02625	0.02599	0.02573	0.02548	0.02522	0.02498
3.7	0.02472	0.02448	0.02423	0.02399	0.02375	0.02352	0.02328	0.02305	0.02282	0.02260
3.8	0.02237	0.02215	0.02193	0.02171	0.02149	0.02128	0.02107	0.02086	0.02065	0.02045
3.9	0.02024	0.02004	0.01984	0.01964	0.01945	0.01925	0.01906	0.01887	0.01869	0.01850
4.0	0.01832	0.01813	0.01795	0.01777	0.01760	0.01742	0.01725	0.01708	0.01691	0.01674
4.1	0.01657	0.01641	0.01624	0.01608	0.01592	0.01576	0.01561	0.01545	0.01530	0.01515
4.2	0.01500	0.01485	0.01470	0.01455	0.01441	0.01426	0.01412	0.01398	0.01384	0.01370
4.3	0.01357	0.01343	0.01330	0.01317	0.01304	0.01291	0.01278	0.01265	0.01253	0.01240
4.4	0.01228	0.01215	0.01203	0.01191	0.01180	0.01168	0.01156	0.01145	0.01133	0.01122
4.5	0.01111	0.01100	0.01089	0.01078	0.01067	0.01057	0.01046	0.01036	0.01025	0.01015
4.6	0.01005	0.00995	0.00985	0.00975	0.00966	0.00956	0.00947	0.00937	0.00928	0.00919
4.7	0.00909	0.00900	0.00891	0.00883	0.00874	0.00865	0.00857	0.00848	0.00840	0.00831
4.8	0.00823	0.00815	0.00807	0.00799	0.00791	0.00783	0.00775	0.00767	0.00760	0.00752
4.9	0.00745	0.00737	0.00730	0.00723	0.00715	0.00708	0.00701	0.00694	0.00687	0.00681
5.0	0.00674	0.00667	0.00660	0.00654	0.00647	0.00641	0.00635	0.00628	0.00622	0.00616

F	0.0[b]	0.1	0.2	0.3	0.4	0.5	0.6	0.7	0.8	0.9
6.0	0.00248	0.00224	0.00203	0.00184	0.00166	0.00150	0.00136	0.00123	0.00111	0.00101
7.0	0.00091	0.00083	0.00075	0.00068	0.00061	0.00055	0.00050	0.00045	0.00041	0.00037
8.0	0.00034	0.00030	0.00027	0.00025	0.00022	0.00020	0.00018	0.00016	0.00015	0.00014
9.0	0.00012	0.00011	0.00010	0.00009	0.00008	0.00007	0.00007	0.00006	0.00006	0.00005

[a] Column headings provide second decimal place for value of F.
[b] Column headings provide first decimal place for value of F.

INDEX

INDEX

Accelerated testing, 105–111
   magnified loading, 108
   sudden death, 109
Adequate performance, 5, 124, 146
Algebra of normal distributions, 243–244
A posteriori probability, 17, 18, 213
Applied loading, 118
A priori probability, 12, 17, 18, 22, 213
Availability, 7, 156–161

Bayes' theorem, 66–69, 213
Binomial distribution, 38–40, 50, 70, 221
Block diagram, 38
Break-in period, 12
Burn-in period, 12, 23, 94

Catastrophic failure, 13
Chance failure, 13
Characteristic life, 102, 245
Coherent system, 132
Combinations, 215–216
Combined systems:
   dual function, 63–66
   series-parallel, 36–38
Conditional probability, 16, 17, 63, 212
Cost analyses, 126

Debugging, 12, 23
De-energized, 32, 34, 54
Derating, 79–84, 172
Design:
   checklist, 121–122

INDEX

Design:
  guidelines, 146
  optimization, 145
  review, 123
Distribution, parameters of: central tendency, 220
  expectation, 220
  mean, 220
  median, 219
  mode, 219
  standard deviation, 220
  variance, 220
Distribution functions:
  cumulative distribution function (cdf), 218
  probability function (point probability function), 218
  probability density function (pdf), 218

Early life, 12–13, 23–25, 94
Economic factors, 126
Energized, 32, 34, 55
Environment, 118
Erlang distribution, 222
Exponential (negative) distribution, 18, 222
  values of, 258–263

Failure modes effects analysis (FMEA), 124, 149
Failure rate:
  basic, 79
  estimated, 139
  generic, 79
  instantaneous, 9, 15–16
  nominal, 79

Gamma distribution, 221
Gaussian distribution (*see* Normal distribution)
General reliability function, 10–11, 14, 25

Hazard rate, 9
Human factors, 122, 144, 162–166

Log normal distribution, 20, 222
Logic diagram, 38

Maintainability, 7, 93, 159
Maintenance, 84–89
  imperfect, 84
  perfect, 84
  preventive, 7, 84
  repair, 7, 84
  time constant (MTC), 159
Maximum likelihood estimator, 103–104
Mean time between failures (MTBF), 16–17, 19, 31, 35, 57, 86, 104, 159
Mean time to failure (MTTF), 16, 31, 53, 86
Mean time to repair (MTTR), 159, 161
Mean wearout life (T_M), 19
Minimal cut, 132
Mortality, 12–13, 15, 34, 55
  figure, 13
Multimode function, 74–77
Multinomial distribution, 42–44, 50

Normal distribution, 20, 222
  algebra of, 243–244
  values of, 257

Parallel systems, 29, 34–36
Parameter estimation, 102–104
Pascal's triangle, 216
Permutations, 75, 215
Poisson distribution, 41, 50, 105, 160
Probability:
  addition law, 212
  a posteriori, 17, 18, 66, 213
  a priori, 12, 17, 18, 22, 66, 213
  Bayes' theorem, 66–69, 213
  conditional, 16, 17, 66, 212
  defined, 8, 209–210
  multiplication law, 66, 212
  permutations, 75, 215
  role of, 5, 210
  unconditional, 16
Probability distributions:
  binomial, 38–40, 50, 70, 221
  Erlang, 222
  exponential (negative), 18, 222
    values of, 258–263
  gamma, 221
  Gaussian, 222
  log normal, 20, 222

Probability distributions:
   multinomial, 42–44, 50
   normal, 20, 222
      algebra of, 243–244
      values of, 257
   Poisson, 41, 50, 105, 160, 221
   Weibull, 23, 26, 107, 112, 222, 239, 245–256

Quadding, 171
   example 4-9, 42
Quiescent, 33

Random failure, 13
Random variables:
   continuous, 217
   discrete, 217
Ranking, 236–242
Redundancy:
   allocation, 188
   parallel, 6, 36, 50
   standby, 7, 50–59, 160
   task, 165
   techniques, 168
Reliability:
   allocation, 181–195
   block diagram, 38
   definition, 3, 4
   demonstrated, 5
   general function, 10–11, 14, 29
   inherent, 4
   model, 148
   predicted, 4
   tasks, 6–7
   testing, 92–102, 227–235
      development and demonstration, 93
      operation, 93
      performance criteria, 100
      qualification and acceptance, 93

Safety factor, 172–174
Series system, 29, 31–32
Sequential testing, 111
Shakedown, 12
Spare switching, 169

Specifications, 122
Standby systems, 50–59
   multiunit, 55–57
   two unit, 52–55
Statistics, definition, 209
Sudden-death testing, 109
Super-reliability, 167–171
System:
   checks, 87
   dual function, 63–66
   failures, 87
   inspections, 87
   parallel, 29, 34–36
   series, 29, 31–32
   series-parallel, 36–38
   standby, 50–59

Testing:
   accelerated, 105
   censored item, 107
   conditions, 98
   nonreplacement, 107
   procedures, 98
   replacement, 107
   sequential, 111
   sudden death, 109
   types: evaluation, 95
      production, 95
      qualification, 95
      service, 95
Tolerances, 122
Tree diagram, 124, 133

Unconditional probability, 16
Unreliability, probability of failure, 9, 18, 29
Useful life, 13, 18–20, 94

Value analysis, 126
Voting circuits, 170

Wearout life, 13, 20–23, 94
Weibull distribution, 23, 26, 107, 112, 222, 239, 245–256
Worst case approach, 118, 172